GENETIC ALGORITHMS AND ROBOTICS
A Heuristic Strategy for Optimization

WORLD SCIENTIFIC SERIES IN ROBOTICS AND AUTOMATED SYSTEMS

Editor-in-Charge: Prof T M Husband
(*Vice Chancellor, University of Salford*)

Vol. 1: Genetic Algorithms and Robotics — A Heuristic Strategy for Optimization (*Y Davidor*)

Forthcoming volumes:

Computer Vision, Models and Inspection (*A D Marshall and R R Martin*)
Intelligent Control (*C J Harris*)
Parallel Computer Systems for Robotics (Eds. *A Bejczy and A Fijany*)
Intelligent Robotic Planning Systems (*P C-Y Sheu*)

World Scientific Series in Robotics and Automated Systems — Vol. 1

GENETIC ALGORITHMS AND ROBOTICS
A Heuristic Strategy for Optimization

YUVAL DAVIDOR

Department of Applied Mathematics and Computer Science
The Weizmann Institute of Science
Rehovot, Israel

World Scientific
Singapore • New Jersey • London • Hong Kong

Published by

World Scientific Publishing Co. Pte. Ltd.
P O Box 128, Farrer Road, Singapore 9128
USA office: 687 Hartwell Street, Teaneck, NJ 07666
UK office: 73 Lynton Mead, Totteridge, London N20 8DH

Library of Congress Cataloging-in-Publication Data

Davidor, Yuval.
    Genetic algorithms and robotics : a heuristic strategy for
optimization / Yuval Davidor.
        p.   cm. –– (World Scientific series in robotics and automated
systems ; vol. 1)
    Includes bibliographical references and index.
    ISBN 9810202172
    1. Robots––Control systems.   2. Combinatorial optimization.
3. Algorithms.   I. Title.   II. Series.
TJ211.35.D38   1990
629.8'92 ––dc20   90–47438
                                                                        CIP

Cover design by Einat Delman.

Printed in Singapore by JBW Printers and Binders Pte. Ltd.

FOREWORD

This series of books and monographs sets out to capture the essence of the current state-of-the-art in Robotics and Automation. Dr Yuval Davidor's book represents an excellent example of this. His book explains, analyses and assesses a novel approach to robotic control which is not only important, but fascinating.

Control strategies for robotic devices based on classical optimization methods are known to be inadequate for important aspects of automated operations. Yet intelligent control is an essential building block of the complex systems constituting advanced robotic installations. Whether robots are used in factory locations for non-trivial tasks or in non-factory, but hazardous environments such as mining, tunnelling or fire fighting, the single major problem facing system designers lies in control engineering. There is a clear need for control strategies which are flexible, but sufficiently 'intelligent' to guide and monitor complex interactions of remotely controlled mechanisms.

Dr Davidor argues convincingly that genetic algorithms offer the scope to help achieve such flexibility in the face of complexity. His book explains the fundamentals underlying genetic algorithms in terminology readily understood by the scientist, engineer or mathematician. He develops his theme to show that existing applications can be extended by using genetic algorithms to generate robot trajectories. He goes on to develop a genetic algorithmic approach which is capable of specifying near optimum trajectories.

Genetic algorithms as a topic of study is set to grow rapidly, and there is a rapidly expanding research community drawn from mathematicians, computer scientists, engineers and biologists around the world. The research community focussing on robotic control engineering is already large and growing fast. Dr Davidor's book has much to offer both communities. His novel and original analyses of a highly interdisciplinary field of investigation seems likely to establish a valuable foundation for new, strategic and effective methodologies.

Professor T. M. Husband
Imperial College
London

PREFACE

This book grew out of research activity I conducted at the Centre for Robotics and the Department of Computing at Imperial College, London, during the years 1986-89, for my doctoral dissertation. While working on process control problems in robotics, I stumbled on a problem which classical optimization techniques could not adequately solve – the optimization of redundant and under-specified systems. I seemed to be stuck. One evening, a phone call from a friend drew my attention to a television programme in which the person interviewed spoke about the type of problems I was trying to solve. I caught the second half of the BBC Horizon episode entitled: *The Blind Watchmaker*. The person interviewed was Professor John H. Holland of the University of Michigan. I obtained the address of Professor Holland from the BBC and wrote to him, only to discover later that he is the 'father' of what is called *genetic algorithms* – a model of adaptation in natural and artificial systems. A model of adaptation which not only offers a powerful optimizing leverage with which a diverse range of engineering problems can be processed, but also sheds new light on the intriguing mechanisms of evolution. Genetic algorithms have been applied to a diverse range of problems. Problems which range from real world applications such as flow control of a gas pipeline, design of airfoil profiles, robot trajectory planning and electronics, to more theoretical problems of combinatorics, game theory, economics and machine learning. For me, the television programme has marked the beginning of an enchanting research odyssey into the secrets of natural adaptation and the mechanisms of evolution – a research odyssey which still continues.

The material covered in this book is interdisciplinary in nature. It combines topics from population genetics, control theory, manufacturing technology, and aspects of computer science. It does this by applying a genetic algorithm (GA), a heuristic probabilistic search procedure, to the domain of robotics. Rather than reproduce good work already available in the literature on the essentials of the above subjects, this book intends to give a detailed description of how to apply a GA to real world problems as it has not thus far received proper treatment. I have chosen as a model the generation of robot trajectories. The choice of the specific problem was motivated by the fact that trajectory generation applies to many process control issues in manufacturing. By applying a GA to a real world problem, such as the generation and optimization of robot trajectories, considerable insight is acquired into the workings of GAs and into the aspects of the human interface with complex domains. It is with this intention and in this spirit that this book is presented – to attempt to provide a clear introduction to the workings of GAs in the context of optimization of large, complex and redundant systems. It is only as a means of introducing and treating the aspects mentioned above that this book tries to provide the basic tools for acquiring an intuitive understanding as to what GAs are, how they work and when it might be rewarding to use them.

The book is divided into two parts. Part I describes GAs and constitutes an illustrative introduction to the art and practice of these special search procedures. It also attempts to provide an answer to that in genetic algorithms which brings computer procedures and nature so intimately together. Part II gives a detailed description of an application of a GA model in the generation and optimization domain of robot trajectories. There are three chapters at the end of Part II which are devoted to some general aspects of learning and adaptation. The subjects of these three chapters emerged from the GA model and the considerable similarity between the model and natural phenomena.

ACKNOWLEDGEMENTS

While enjoying myself with the research, I got to know people who not only helped me with my work, but who also impressed me with their personality. Some of them, as often happens, may not even be aware of the extent of influence they had on me, and the book. Their extensive contribution to the book, to my knowledge and understanding of the subject, is most appreciated. To them, and to their unstinted interest in research, I owe a sincere gratitude. I wish to thank the Centre for Robotics and Automated Systems, and the Department of Computing at Imperial College, where most of this work was carried out.

I particularly want to thank Antonia Jones, of the Department of Computing at Imperial College, who served as one of my dissertation supervisors, taught me scientific discipline, and tried to educate me in the art of coherent writing. I enjoyed her unconditional support and benefitted greatly from her patience and knowledge. I thank Tom Westerdale, of Birkbeck College, who was a well of unpublished history of GAs and a wall against which I could bounce ideas. I thank Tom Husband, the Head of the Department of Mechanical Engineering at Imperial College, whose support I enjoyed throughout and who enabled me to ride some tempestuous incidents during my stay at the Centre for Robotics and Automated Systems. I also want to thank David Goldberg, of the University of Alabama, for his ongoing support and advice extended wholeheartedly over the last couple of years, and for comments which helped me improve the quality of my scientific work.

I want to express special thanks to Leon Zlajpah of Jozef Stefan Institute, Yugoslavia, for allowing me the free use of a robot dynamic simulation he developed, and for active support in tailoring the simulation to suit the special requirements set by this work.

Finally, I wish to thank Einat Delman who had to put up with my capriciousness while designing the cover for this book.

CONTENTS

PART I

THE GENETIC ALGORITHMS
PHILOSOPHY

Chapter 1
YET ANOTHER SEARCH METHOD

Robot systems are an example of systems that are the result of the integration of a medley of subsystems and feedback loops where literally hundreds of variables may affect their operation. The core problem in robot process control, and also a frequent problem in many other systems, is that the optimal values of the system's control parameters are not known, and there is no straightforward algorithm to discover them. Traditional optimization techniques, such as self tuning and adaptive control, depend too greatly on a deterministic relationship between the control parameters and the resulting performance; an explicit effect of an input signal on the performance of the system. These techniques have improved, but are unable to optimize the performance of very complex systems.

Attempts have been made (in the manufacturing environment) to overcome the effect of this plethora of parameters influencing the performance by recording all the relevant data, and developing procedures to handle and interpret these data. These attempts have avoided the main issue as the programmer of these control procedures is still required to possess a thorough understanding of the system's basic concepts; an understanding which he often does not have, nor has the time to acquire. The success, which conventional procedures do achieve, is either in situations where the system can be modelled with sufficient accuracy, or when the number of plausible parameter values is small enough to be tested exhaustively. For many systems that cannot be adequately modelled, or whose state space is too large, a different technique is required – one that will efficiently search for the system's optimal control parameter values without having to rely on prior knowledge of the performance space.

The model presented in this book is an extension of previous studies of how evolution-like mechanisms can direct and improve an automated process without requiring precise control of the environment. Through the ages, developments and solutions for problems have been found by copying solutions to similar problems found in nature. The number of such 'copyright' violations is staggering. The list of scientific developments initiated by observing nature is long, and the search for nature-like solutions to complex problems continues. Engineers resort to nature and to natural mechanisms to shed light on situations too complicated to handle with any of the classical scientific tools. Those more reluctant to make the analogy between the natural and the artificial, turn to computer models to better understand natural phenomena. Our discussion throughout this book is based on the following central dogma: in nature, species are well adapted to their environmental niche in spite of the organisms' great complexity. Noting the richness of the genetic pool organisms already possess, we come to the conclusion that the number of organisms which are present is nothing but a minuscule percentage of the possible diversity nature can provide. Therefore, the high adaptation species exhibit was obtained by investigating only a relatively small number of plausible genetic solutions. The complexity of many manufacturing systems resembles the complexity found in the natural environment. Thus, in the attempt to overcome prohibitive complexity one may draw analogies from nature and learn from its search mechanisms, *i.e.* robust mechanisms that efficiently search for sets of co-adapted characteristics capable of improving the overall fitness of the species.

'Search' in this book means a process of locating a particular solution to a given problem among a finite number of plausible solutions. A solution is said to solve the problem if it satisfies a given objective function. The set of plausible solutions is called the solution space. The object of a search procedure is to minimize the number of objective function evaluations necessary to locate a satisfactory solution. A search procedure is said to be efficient if the number of solutions evaluated is small in comparison to the size of the search space. The smaller this ratio, the more efficient the search procedure is. Robustness is a measure of how efficiency of a given search procedure changes drastically when the problem parameters are changed slightly. A certain search procedure may be very efficient in solving a specific problem, but if this efficiency decreases drastically for other problems, this search procedure is not a robust procedure. The different search procedures are divided into three main search categories: calculus-based, enumerative and heuristic. Calculus-based procedures use either analytical or numerical models of the solution space as the basis for the search. Enumerative procedures search systematically and do not incorporate any sophisticated mechanisms. The heuristic procedures attempt to improve on the search efficiency of the enumerative methods without incorporating models of the solution space which are often unavailable.

This chapter introduces a search procedure which joins the existing diverse family of search procedures. Before outlining the procedure, three popular problems representing typical problems in control, optimization, and engineering, are presented by way of illustration. These problems cannot be solved efficiently and robustly by a straightforward algorithm. Although they appear to be unrelated to each other, they possess common features that can be realized and utilized by a 'blind' heuristic search strategy called *genetic algorithms* – a general computation procedure for search and optimization based on simplified models of population genetics. This computation model, its mechanisms and their effect on adaptation and search efficiency, shall be shown hereafter.

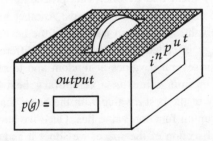

Fig. 1 - *The coin-operated black box paradigm.*

An Assemblage of Difficult Problems

It is instructive to learn which aspects of a problem make it difficult for traditional search procedures, but are less of a problem to GAs. This section aims, therefore, at presenting the two aspects of search procedures, efficiency and robustness, in three illustrative examples.

Function Optimization and the Black Box Paradigm

Function optimization is a good example by which to examine the robustness of search procedures, and the frequent trade-off between robustness and efficiency. Robustness of search procedures is presented in the following way. You are given a coin-operated black box with an input port and an output display. The black box contains an unknown function, $f(d): R \rightarrow R$. The operating instructions supplied with the black box indicate that upon receiving a penny token, the box takes the operand, d, from the input port, calculates its function value, $f(d)$, and displays this value on the output display (see Fig. 1 for illustration). The black box does not perform any other operations, and no clues, such as continuity, differentiability, etc., are given as to the nature of the function inside the black box. You are requested to find the number d (without loss of generality $\{0 \leq d \leq 1\}$), which corresponds to the maximum function value in the given interval, while spending the minimum amount of money. Let us consider the options. One possible search strategy would be to list all possible d's and to try them one by one. This procedure is called exhaustive search because of the effect it has on our funds. Another method might be to simply try different d values at random and record the best one we find as we go along. This strategy is called random search. A different search strategy might be to pick again several d values at random and to choose the one that yields the largest function value as a temporary best bet, and then to investigate the vicinity of the best d value with the belief that this is the region where the global maximum function value lies. Once we have found a better d value in this narrowed section of the space, we adopt it and continue to search again in its vicinity. There are many other possible search strategies, some simple, others quite sophisticated.

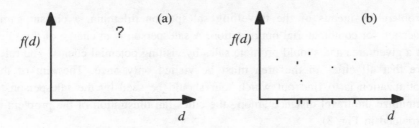

Fig. 2 - *The black box (a), and the discrete sampling (b).*

Fig. 2a illustrates our lack of knowledge about the nature of the function in the black box, and Fig. 2b illustrates the search process by means of the black box (providing an input from the domain and observing its range displayed at the output). What strategy would you choose, bearing in mind the limited information supplied by the black box on the nature of the function, and that the black box requires a penny for every function evaluation?

The difficulty of constructing an efficient search procedure to solve an unknown function is further intensified when a different version of the black box is introduced. In the new model, a door is added to the box so that the function it contains can be replaced, at will, by different functions. Further, we take the black box to the classroom and divide the students into two non-communicating groups. One group is given free rein to devise functions over the set $\mathcal{D}$, while the second group of students is challenged with the task of constructing an algorithm which shall require the minimum search resources over the entire set of functions proposed by the first group of students (remember that one penny is being charged for every function evaluation). In other words, we require an algorithm that on average over an arbitrary set of functions will require the minimum search resources.

The Travelling Salesperson Problem

The first illustrated problem demonstrated the robustness required from a search procedure which is intended to be used in complex domains. We now consider another example to illustrate the problem of searching large spaces. Large solution spaces exist in numerous real world problems. They pose a serious technical problem, as excessively large search spaces require a very efficient search procedure regardless of robustness. Let us illustrate the problem by means of the travelling salesperson dilemma, a classic game question for combinatorial optimization. A salesperson is in charge of the sales in a given area and should promote sales by visiting potential clients. The rules are that all cities in the area must be visited only once. The aim of the optimization is to find out which tour should be used by the salesperson to minimize the travel distance among the cities (an illustration of the problem is provided in Fig. 3).

The salesperson always starts the tour from the same city (his own), and then chooses a sequence of visits which in his opinion require the minimum travel distance. The TSP can also be stated in the following way: Given a graph of n nodes, what is the shortest Hamiltonian path connecting the n nodes? The number of alternative routes is very large. We are interested in developing an algorithm that can realize such a tour within a reasonable time space.

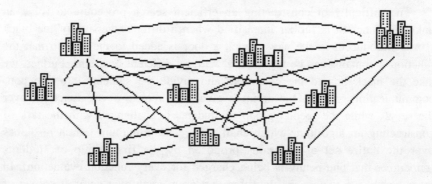

Fig. 3 - An illustration of a nine city TSP.

Robot Trajectory Generation

The third example presented here is a real world application – the optimization of robot trajectories. Most robot applications are based on a motion trajectory composed of a sequence of spatial displacements of the robot arm. Mechanically, a robot arm is an open kinematic chain comprising relatively stiff links with a joint between adjacent links. Each link represents one degree of freedom and can be commanded to move independently of all other links.

Standard systems have six degrees of freedom spatially configurated, to obtain full spatial flexibility. Since a robot arm performs a task through the motion of its end-effector attached to the last link, the last link is the primary component of the whole structure and its loose end is called the *end-effector* (Fig. 4).

An end-effector trajectory is created by programming a sequence of end-effector positions for the arm to follow. Because only a finite number of positions can be specified and stored in any given program, the robot motion controller has to move the arm between the discrete positions following a dead reckoning procedure. As a result of the arm motion between discrete positions, the end-effector draws a continuous path. Optimization of robot trajectories means the identification of the optimum combination and number of intermediate positions, and that means a great many alternative trajectories.

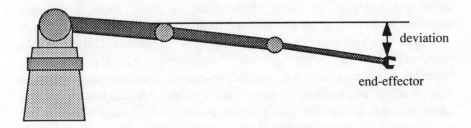

Fig. 4 - *3-link planar robot arm at a fully stretched position. The end-effector exhibits a steady-state positioning deviation from the horizontal due to limited positioning accuracy.*

The complexity of programming a trajectory can be appreciated by examining the vertical plane in which the end-effector is required to follow the straight line connecting points A and B (Fig. 5a). One robot program can list sites 1 and 2 for the end-effector to visit (Fig. 5b) while another program considers sites 3, 4, 5 and 6 to be more accurate (Fig. 5c). Yet another possible trajectory, which also accounts for the effect of gravity, is composed of sites 7, 8, 9 and 10 (Fig. 5d).

The quality of the resulting end-effector path of each of the alternative programs is quite different, though all trajectories aim at taking the arm from A to B as closely as possible to the straight line connecting the two points. The theoretically equivocal question, how many intermediate positions are necessary, is further complicated by the fact that intermediate positions

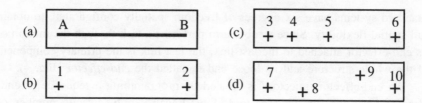

Fig. 5 - *Illustrating the complexity in generating a trajectory. (a) The desired straight line path between the starting end A and the terminating end B. (b) One possible trajectory specified by two arm-configurations that correspond to the end-effector being at the two ends of the desired path. (c) Another alternative trajectory with four end-effector positions along the desired path. (d) Another possible trajectory specification for the straight line path.*

have no clear structure and two trajectories have no clear homology in their defining end-effector positions. Furthermore, too often, robot systems are assumed to have a predetermined performance. The ambiguity between specified and performed paths turns trajectory generation into a highly nonlinear problem with a very large state space. Although the above descriptions of the system are minimal, it is possible to see that mechanical inaccuracies, mechanical coupling and dynamic effects turn trajectory generation into a very complex problem to optimize. The following section will present a simple search procedure which will later be shown as possessing the potential to address these problems as well as other similar ones.

The three problems which were presented are really an archetype of many complex problems, problems which are not fully defined, problems which have many alternative solutions which are impractical to evaluate systematically. To this category also belong problems whose specifications may change in time, and those which may incorporate a substantial degree of redundancy. All these problems are difficult, and present substantial challenges for traditional optimizing techniques. Let us now analyze a novel search procedure which has the potential of improving on the shortcomings of traditional search procedures.

A Blind Search Strategy

A natural strategy one would tend to use in complex solution spaces, where there is no straightforward algorithm to search for the optimum solution, is to try a few guesses and to focus on small regions where further search efforts seem likely to show promise. The decision as to where to invest search efforts is usually based on local progress of previous guesses. However, in complex domains this usually proves to be inadequate. Let us therefore consider a different search strategy.

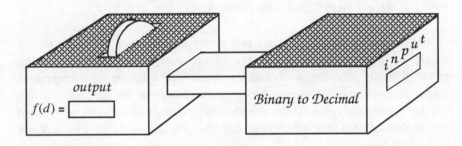

Fig. 6 - *The binary to decimal converter attached to the black box.*

Returning to the black box, to which this time we add an additional box that converts unsigned integer binary numbers l bits long to real numbers (Fig. 6). The black box itself as described by Fig. 1 is not altered. One simply uses unsigned integer binary representation to represent the real numbers. The way the binary to decimal converter operates is described by the following function $p(g)$:

$$d = p(g) = \sum_{i=1}^{l} g_i 2^{-i} . \qquad (1)$$

For example, the binary string of length 5, 01101, represents the real number 0.40625 because

$$0*1^{-1} + 1*2^{-2} + 1*2^{-3} + 0*2^{-4} + 1*2^{-5} = 0.40625.$$

We choose at random a sample of binary strings, say 100 strings, and use the modified black box to evaluate them (note that from now on we only manipulate a binary representation of the arguments d). We attach the function value (the output of the black box) to the corresponding binary string, so that each binary string has its function value associated with it. We are now ready to start the search for strings which correspond to higher function values. The search process progresses according to the following scheme:

Step 1 – Choose a member of the sample <u>probabilistically</u> according to its function value.

Step 2 – Duplicate the chosen binary string and put the duplicant aside.

Step 3 – Repeat steps 1 and 2 one more time.

At the end of step 3, we have two binary strings which are duplicants of binary strings we already possess in the sample. Choosing a string probabilistically according to its function value means that on the average (one cannot be more specific due to the stochastic nature of the process) a string that has a function value 10 times higher than another string shall be selected and duplicated 10 times more frequently than the one with the low function value. We then:

Step 4 – Choose at <u>random</u> a position along the binary string and mark this position on the two duplicants as cross site 1.

Step 5 – Choose at <u>random</u> a second position along the binary string and mark this position on the two duplicants as cross site 2.

Step 6 – Crossover counter-segments between the marked cross sites (Fig. 7).

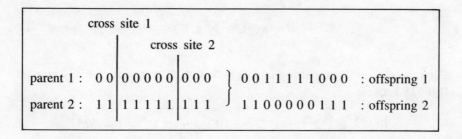

Fig. 7 – An illustrative crossover operation with 10-bit binary strings.

At the end of step 6 we obtained two new binary strings which we shall call *offspring* (although it is possible that one or both offspring match identically members in the sample, we shall regard them as 'new' for reasons which shall become clear later). We proceed:

Step 7 – Evaluate the function value of the two offspring binary strings through the black box.

Step 8 – Choose at random two members of the sample and replace them with the two offspring.

Step 9 – Repeat from step 1.

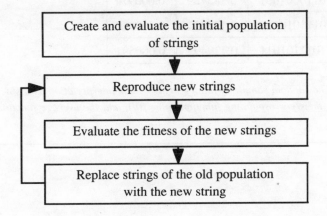

Fig. 8 - *An outline for a genetic algorithm.*

What we have just described is a very simple model of evolution incorporating natural selection, survival of the fittest and sexual reproduction (summarized in Fig. 8). Before suggesting an explanation as to why it is asserted that in certain problem domains this search procedure is better than other search procedures, we shall use the black box paradigm one more time to actually process an optimization of an unknown function given to us in the black box.

Illustrated Use of the Black Box

We shall begin the search by going through each of the sequential steps so we can follow its mechanisms. Thereafter, we shall allow the computer to take over in order to speed-up the process. We choose strings of length 10 and the

No.	string	d	$f(d)$
1	0110100000	0.406250	0.037941
2	1101000011	0.815430	0.069108
3	1101101110	0.857422	0.174605
4	0001100101	0.098633	0.363379
5	1100010001	0.766602	0.000021
6	0110001001	0.383789	0.000974
7	0111010100	0.457031	0.399710
8	1011101011	0.729492	0.000591

$\overline{f(d)}$
0.130791

Table 1 - *The random sample of binary strings, their interpretation as a single decision variable d, their corresponding function value f(d), and the average function value.*

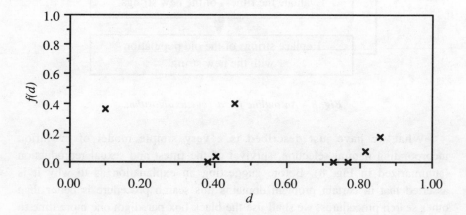

Fig. 9 - *The first 8 evaluations through the black box.*

population size 8 (these values are selected purely for illustrative considerations, and in a real application of GAs they shall preferably be much larger). A black box is given to us with an unknown function inside. We choose a sample population of strings at <u>random</u> (the sample population is summarized in Table 1 and also plotted in Fig. 9).

The two random numbers, 9.81 and 66.74, select <u>probabilistically</u> strings 2 and 7 (the probabilities for reproduction are given in Fig. 10 and Table 1). The two cross sites 1 and 6 (picked at <u>random</u>) conclude the reproduction of two new offspring strings (Fig. 11).

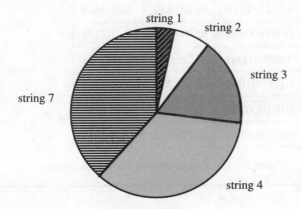

Fig. 10 - *A pie chart of the probability distribution for selecting a string. Notice that strings 5, 6 and 8 have a negligible probability (from Table 1).*

1	10100	0011	1111010011
		⇒	
0	11101	0100	0101000100

Fig. 11 - *The parent strings, the two random cross sites, and the resulting offspring strings.*

The fitness of the two offspring strings is evaluated and they then replace two strings of the sample which are selected at <u>random</u> (Table 2).

It is clear that such a small sample size is too prone to influence by sampling error. We shall, therefore, increase the length of the strings to $l=30$ (resulting in a search size of $2^{30} \approx 10^9$) and the population size to 50, and let the computer follow steps 1 through 9 with the black box. Fig. 12 plots the initial sample, Fig. 13 presents the population after 40 generations while Fig. 14 presents the population after generation 100.

No.	string	d	$f(d)$
1	0110100000	0.406250	0.037941
9	1111010011	0.956055	0.000000
3	1101101110	0.857422	0.174605
4	0001100101	0.098633	0.363379
5	1100010001	0.766602	0.000021
6	0110001001	0.383789	0.000974
10	0101000100	0.316406	0.041796
8	1011101011	0.729492	0.000591

$\overline{f(d)}$
0.077413

Table 2 *- The sample after replacing two strings.*

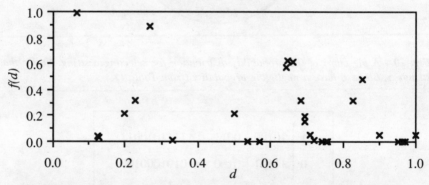

Fig. 12 *- The initial sample of search points coded with 30-bit binary strings.*

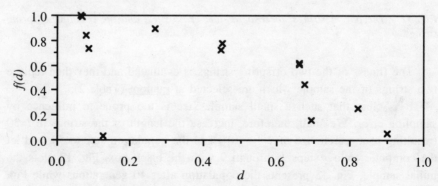

Fig. 13 *- The sample after 40 iterations.*

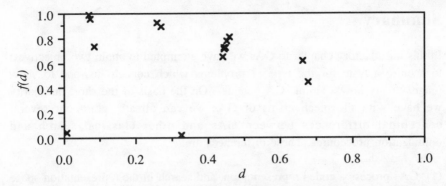

Fig. 14 *- The sample after 100 iterations.*

Finally, we open the black box and investigate the function which was hidden inside (a small yet didactical violation of the black box operating instructions). Fig. 15 plots the function which was in the black box (the function was adapted from [Goldberg and Richardson, 1987]). Now that we know how the function behaves, we can appreciate the adaptive quality of the genetic search. There is a clear migration of 'x' upwards and towards the left. It is as if the 'x' realized that it is much better to climb local peaks and also that higher peaks are found to the 'west'. It is true that we have not appreciably improved the best result because we were 'lucky' to choose a very good point in the initial population (the left-most 'x' in Fig. 12). However, because we never know in advance how good our best guesses are, we have an algorithm that adapts itself and confirms its findings.

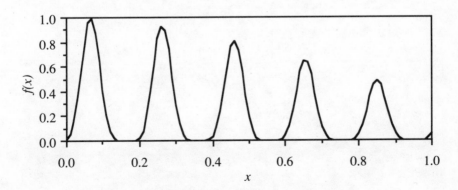

Fig. 15 *- The function f(d) that was inside the black box.*

Summary

In this introductory chapter to GAs we have attempted to attain two objectives: to acquire a taste for the type of problems which concern us, and to show pragmatically how a simple GA operates. On the basis of the short familiarity we have with the mechanisms of GAs we can already point out several conceptual differences between GAs and other classical search and optimization procedures. These differences are:

(1) GAs process a coded representation, and search in the representation space rather than directly in the original domain.
(2) GAs require a population of solutions instead of a single solution.
(3) GAs use global information from the entire space.
(4) GAs use 'blind' transition operators and not deterministic ones.
(5) GAs do not require domain-specific information.

In the following chapter we shall analyze GAs in greater detail and formulate their behaviour. GAs will be portrayed as a search strategy which contains some distinct differences in relation to other classical search strategies, differences which possess an advantage in certain complex domains such as the three problems we have illustrated at the opening of this chapter.

One comment regarding the illustrated operation of GAs (pages 9 through 13). The number of offspring generated at each generation may not necessarily be 2 as suggested, and usually will be equal to the population size.

Chapter 2
WHY GO GENETIC?

In this chapter, the genetic strategy for search and adaptation will be reviewed in a disciplined and careful fashion. By the end of this chapter, the reader should have gained some notion of what happens in nature and how evolutionists explain the impressive phenomenon of natural adaptation. A simple model of a GA will be proposed, and its mechanisms will be explained. Finally we shall attempt to explain the special features of GAs with the schema theorem, the fundamental theorem of GAs.

Evolutionary Mechanisms as Adaptive Search Procedures[1]

Sociobiology emerged from the attempt to explain altruism – the willingness of some animals to risk their own life in an effort to protect their offspring – a phenomenon which Darwin's doctrine of evolution finds very difficult to

---

[1] The material presented in this section is based on several textbooks, but primarily on a special issue on evolutionary theories presented in *Machshavot*. Vol. 50-51, Oct, 11981, IBM, Israel.

explain. For, if natural selection works against apparently detrimental qualities within the organism, it should have extinguished those traits that make their bearer sacrifice self in order to benefit others. To overcome this old dilemma, sociobiology shifted the level at which natural selection works from the individual to the chromosome level. This means that a bird which risks its life for its young does in fact serve an interest, but one that concerns its genes rather than its personal survival, because the young who carry the parent genes have a better chance to reproduce in the future.

By thus transferring the sphere where natural selection operates, the genes become the basic building blocks of natural selection, placing the organism with its impulses, habits and characteristics, in subordination to its genes. From this viewpoint, the will to live is not a defence mechanism to protect the organism, but rather, a mechanism that improves the chances of the genes to be replicated – even at the price of the organism's life. This also means that qualities like love, hate, compassion and solidarity are traits that have survived the test of natural selection, and lasted, because they served the genes' interest. Human behaviour cannot, therefore, be claimed to be formed solely by culture, especially due to its short existence (only 10,000 years or so). It has to be viewed as a product of genetic evolution that has been working for several million years.

According to Darwin, however, the basis of evolution is the occurrence of random hereditary modifications in the organisms of a species. The advantageous modifications are then adopted and the disadvantageous ones discarded through natural selection. The organism carrying a negative change, a change that results in an inferiority relative to its fellow organisms, will find the struggle for food and a mating partner difficult, and become more vulnerable to the hostile environment. The Darwinian theory of evolution, with the capriciousness, passivity, and mechanical nature of the evolutionary process, supplies to date the best rational explanation of the evolution and the adaptability of species.

However, even if the concept of gene-driven evolution is correct, there are many difficult questions left unanswered. For example: why did natural selection find it beneficial to equip man before he became aware of fire with a potent mind capable of building the nuclear bomb? That is, if a mutation is being preserved due to its functional advantages and its contribution to the present, it is difficult to justify the appearance of a brain so developed for its time with a potential that will be utilized only thousands of years later. It is as strange as if the pre-historic antelope were not only capable of escaping from the carnivorous cheetah, but also from the bullets of the latter day hunter. The fact that the antelope is capable of escaping from the cheetah with which it has been familiar for 500,000 years, but not bullets, supports the principle that natural selection is driven by present rather than future needs. Nevertheless, we should remember that natural selection never ceases to affect evolution, on the macro- as well as the microscopic level. If we accept the concept that all that is around us is a result of selection which is biased towards better fitted mechanisms, then we should look for this advantage even when, on first impression, some mechanisms seem to be less than optimal (such as the immune system that reduces momentary fitness as a result of an immunization memory of diseases which though infrequent have the nature of plagues and tend to re-occur in cycles).

Why Use a Representation Instead of the Real Thing?

It is clear that one neither loses nor gains variety by presenting individual solutions using different representations. Consider the set $\{0,1,2,...,63\}$. An octal representation will use an alphabet of cardinality 8 and therefore requires 2 characters to spread the full set ($8^2=64$), while a binary alphabet representation (which has a cardinality 2), will need 6 characters ($2^6=64$). In whatever way one chooses to represent the space, the full space must be considered. This feature of the representation must be stressed. It is not the solution space one attempts to manipulate by representing it in different ways. Rather, it is the knowledge about the structure of the solution space and the possibility of capturing such knowledge in order to improve it.

Different representations aim at improved identification of similarity features in the solution space. Without conceding loss of variety, we are free to choose any representation we find suitable. The question is which representation format better captures the important features of a given space. It is difficult to foresee what are the possible similarity aspects that exist in a given space. For example, in the one-dimensional real number function space (like the one given in the previous chapter in the black box paradigm), one such possible aspect is the topology of the arguments. All the real numbers in the vicinity of a local maxima may exhibit a correlation between their value and their corresponding function value as long as the function does not change too rapidly. Similarly, if we consider cyclic functions such as the trigonometric functions, we would find that there is a correlation between the Euclidean distances between points of similar function value. In contrast to the above two examples, it is more difficult to discover a structure in the parity problem. In fact, the parity problem has no structure, and its function value space is a monotonous zero with a single fitness spike.

The variety does not change with different representations. What does change, when we represent a space with an auxiliary space, is the perspective and resolution with which one is able to inspect that space. Consider again the binary and octal representations. By knowing that a given string has a '0' as its most significant bit (the first left digit in our discussion), we can already note that this string is in the first half of the solution space. A similar comment can be made if a string has a '1' in the most significant bit, and this follows with all the other digits (6 in total). However, one can make only two such comments with an octal representation because the representation contains only two digits. If one examines the similarity between two or more strings, more comments and information about similarity among the strings can be extracted when the representation is detailed. We therefore reach the conclusion that, everything else being equal, it is advantageous to represent a space in a format in as detailed a manner as possible. The binary representation with its cardinality being 2 is the most detailed representation one can devise for representing real numbers (it should be remembered that we use the binary and real numbers throughout our discussion for purely illustrative purposes and it is by no means intended to suggest that GAs are restricted to processing real numbers through the binary representation). The statement that everything else is equal should be substantiated, and so should the discussion about the gain in adopting a detailed representation. This will be done shortly with the introduction of the schema theorem. In the meantime, our aim is to acquire insight into the representation philosophy which we regard as adopted from nature. Thus, the string representation is analogous to the archetypal information structure suggested by the chromosome.

The game 'Mastermind' is a perfect example for demonstrating the essence of representation. In the game there are two 4-digit natural numbers, one for each of the two players. Each player chooses his 4-digit number and keeps it hidden from the other player, whose aim is to discover that number before the other player discovers his. The only restriction placed on the players in choosing their numbers is that the numbers should not contain repeating digits. Because the order of the digits is important, the number $\mathcal{N}$ of different such 4-digit numbers is, $\mathcal{N}=10*9*8*7=5040$.

The game proceeds when each player in his turn constructs a 4-digit number, and challenges his opponent with this number. The opponent responds by indicating how many digits of the proposed number are correct in themselves, and how many of them are also in their correct position. For example, if the chosen number is 4710 and the first guessed number is 5018, then the response should indicate that there are two correct digits and that one of them is also in its correct position. The players try to find the correct 4 digits and place them in their correct positions by a process of elimination. If the second guessed number is 7921 and therefore has two correct digits, but the third guess 2583 has none, then '1' is a correct digit and its place is not at the far right. Experience has shown that it takes about 6 attempts for good players to guess the opponents number correctly.

However, if we were to modify the representation of the 4-digit numbers so that instead of 4-digit numbers the players were to use an enumerative table with each 4-digit number having a running entry code, then the game would degenerate into a random search and become painfully boring. Although we have neither destroyed the variety of the domain, nor changed any of the rules (the number of correct digits and correct positions is still available to the players), the game loses its essence because the players cannot observe correlations between their guessed numbers and the correctness of the guess. In other words, the players have no way of identifying sub-goals (see Chapter 8 for a discussion on sub-goal reward in machine learning and GAs). Again, the similarity templates used by the player, and the similarity among 4-digit numbers enables experienced players to work out the correct number with only a few guesses out of the large number of possible combinations (5040!). The following section will provide tools to analyze the effect of similarity templates among solutions in a given space, and will show how GAs, a 'blind' probabilistic heuristic search strategy, are in fact a powerful adaptive search strategy for solution spaces that can be decomposed into similarity templates.

The Schemata Processing Paradigm

By accepting a chromosome representation structure and manipulating a population of strings (Fig. 16 gives a schematic summary of the GAs environment) for search in multi-modal, nonlinear spaces, one shifts the search emphasis from the complete strings to the discovery of co-adapted partial strings, or better still, to the process by which co-adapted partial strings are identified. Indeed, if a better understanding of the problem domain is desired, then it is useful to study the string similarities together with their corresponding fitnesses, and especially to investigate structural similarities of exceptionally fit strings.

"...In some sense we are no longer interested in strings as strings alone. Since important similarities among highly fit strings can help guide a search, we question how one string can be similar to its fellow strings. Specifically we ask, in what ways is a string a representative of the other string classes with similarities at certain string positions? The framework of schemata provides the tool to answer these questions." [Goldberg, 1989b, p.19].

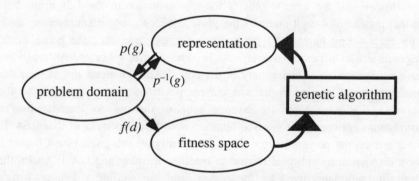

Fig. 16 - *The cycle of functional relationships among the three spaces: representation, domain and fitness, and the manipulation of a GA.*

Formally, we define a similarity template among strings with a *schema* [Holland, 1975]. A schema (over the binary alphabet without loss of generality) is a string of the type

$$(a_1, a_2, ..., a_i, ..., a_l), a_i \in \{0, 1, *\} .$$

The '*' symbol is a 'don't care' symbol which accepts both '1' and '0'. A schema is a template that describes a sub-space of strings that match the schema at all loci where the schema is specific (specifies either '1' or '0'), and regardless of the value the strings exhibit at the loci of the '*' symbol. Therefore, the more '*' symbols a schema contains the less specific it becomes, *i.e.* the more strings it describes. With an alphabet of 2 symbols a schema having m '*' symbols will describe 2^m strings. For example: the schema of length 5 (*,1,*,0,0) describes the set

$$\{(0,1,1,0,0), (1,1,1,0,0), (0,1,0,0,0), (1,1,0,0,0)\},$$

while the schema (*,1,1,0,0) describes the set

$$\{(0,1,1,0,0), (1,1,1,0,0)\}.$$

Usually, by examining the fitness of any one string (whether it be large or small), one might expect to obtain information about other strings which have a structure similar to it. The total number of different schemata resulting from a string of length l, and binary alphabet, is 3^l. The size of the actual search space remains as k^l, since the '*' symbols are not involved in the actual strings GAs process; they only serve as a tool to explain the chromosome structure, and to describe string similarity templates in a compact format. In general, any particular string belongs to exactly 2^l schemata of varying degrees of specificity. The benefits of describing a search space through the framework of schemata can be visualized when string fitness is considered.

The fitness of one particular string contributes little understanding as to how to select subsequent search points (strings), even if the considered string has a relatively high fitness, because it does not indicate which parameters contributed to the high degree of fitness. The only inference one can make is that co-adapted parameters are 'hiding' somewhere in that string. However, when a population of strings and their corresponding fitnesses are considered, ample information can be extracted due to the similarity between strings and their corresponding fitness. Let us remember the mechanism of schemata processing in the game of Mastermind mentioned earlier. The schema theorem does that in a rigorous way.

The mathematical basis of Holland's schema theorem arises from the observation that in evaluating the fitness of a string one also derives implicit knowledge about the schemata which describe that string. The accuracy of this extrapolation depends on the specificity of the given schema. At the micro-level of a GA, the search is viewed through the space of strings. However, the essence of the schema theorem is that one can also view the changing population as a search through the set of schemata which the strings instantiate. Since each string is an instantiation of 2^l possible schemata, in testing a string one derives a great deal of implicit information regarding the 'fitnesses' of the schemata it belongs to. Holland calls this *implicit parallelism* and this observation is a major part of the explanation of the power of the GA search.

With the schema theorem in hand, the essence of the chromosome structure in GA optimization becomes clearer. By selecting strings from a sampled population with a probability relative to their fitnesses, one selects representatives of a particular schemata proportionate to their average fitness. The average fitness of a schema is an artificial quantity that only indicates which string templates are more promising to investigate, and by how much more.

The frequency $m(H, t)$ of a schema H at generation t, will change at generation time $t + 1$ proportionally to the respective selection probability for reproduction. More precisely, the growth of a schema due to the proportional replication is given by:

$$m(H, t + 1) = m(H, t) \frac{\overline{f(H)}}{\overline{f}}, \qquad (2)$$

where the numerator is the average fitness of all strings belonging to the schema H. Similarly, the denominator is the average fitness of the entire population (alternatively one can regard this overall average as the fitness of the schema composed entirely of 'don't care' symbols).

Schemata may be disrupted due to crossover (unless, of course, the crossover is performed between identical strings), and therefore, the expected growth of a schema of Eq. (2) is disrupted accordingly. If P_c is the probability for a crossover, and $\delta(H)$ is the metric distance between the first and last specific schema positions, then the probability of a schema to be disrupted due to crossover is given by

$$P_c \frac{\delta(H)}{l - 1} \qquad (3)$$

and $\delta(H)$ is calculated as shown in Fig. 17.

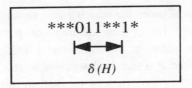

Fig. 17 - *The defining length of a 10-element string, but only four are specific.*

A schema can also be disrupted due to mutation. If the probability of a mutation is P_m and the number of specific positions contained in the schema is denoted by the order of the schema $o(H)$, then the probability of a schema being disrupted by a mutation is given by:

$$o(H)\, P_m, \qquad\qquad (4)$$

and the approximated growth of a schema under crossover and mutation is:

$$m(H, t+1) \geq m(H, t)\, \frac{\overline{f(H)}}{\bar{f}} \left[1 - P_c \frac{\delta(H)}{l-1} - o(H)P_m\right]. \qquad (5)$$

The schema structure as described herein suggests that different schemata may have varying average fitnesses and different specificity. The characteristic feature of schemata leads to another quality of the string space, its inherent sub-division into sub-spaces defined by the schemata. Because schemata are associated with a characteristic fitness, it follows that the sub-search spaces have a characteristic fitness. The relationship between the search space and the performance space is exactly the relationship one wants to establish. Of course, schemata can be partially ordered in a hierarchy of set-theoretic inclusions, but what is more important to the search is the identification of sets of schemata which have above average fitness. Just as one can describe a schema as a 'hyperplane' string so one can imagine sets of schemata as being described by a higher order object, usually called a hyperplane schemata. These hyperplane schemata play an important role in investigating the structure of large, complex spaces.

The schema theorem explains why GAs exhibit high efficiency in search spaces that contain structurally similar sub-spaces, *i.e.* similarities that can be associated with characteristic performance. Thus, by observing the similarities among strings one can advance the search with great efficiency resulting from the implicit parallelism. The schema theorem given in Eq. (5), suggests an exponential efficiency which means that from a single string evaluation we derive information on the 2^l schemata that it instantiates. Although correct, this is not a correct measure of the extent of parallelism of the genetic search both because certain schemata do not survive reproduction, and the number of different schemata which are considered are dependent on the population size. We presented a GA as a search procedure which searches through the space of schemata and we therefore ask how many schemata are being processed at each evaluation cycle. Analyzing schema survival, Holland estimated the number of schemata processed at each evaluation generation in a population of n strings. His estimate is $o(n^3)$, proportional to the cube of the number, n, of strings being processed [Goldberg, 1985; Holland, 1975]. This is a more conservative figure, but nevertheless impressive in its processing power.

In order to capitalize on the advantages of a search via processing schemata, the schemata should be as detailed as possible so as to increase the search efficiency. The importance of a rich structure is one of the main prerequisites the schema theorem requires. Holland refers to this effect, and comments:

"...This suggests that, for adaptive plans which can use the increased information flow (such as the reproductive plans), many detectors deciding among few attributes are preferable to few detectors with a range of many attributes. In genetics this would correspond to chromosomes with many loci and few alleles per locus (the usual case) rather than few loci and many alleles per locus." [Holland, 1975, p. 74].

Further, the schema theorem indicates some necessary conditions for utilizing the implicit parallelism. In order to give the schemata processing its maximum leverage, the disruption of above average fitness schemata should be minimized (the terms in the square brackets in Eq. 4). The disruptive arguments in Eq. 4 are the probability of disrupting a schema due to crossover and mutation. Four arguments influence the disruption: the probability for crossover, P_c, the defining length of a schema, $\delta(H)$, the probability for a mutation, P_m, and the order of a schema, $o(H)$. There is a sufficient body of work that shows that 'tuning' the probabilities P_c and P_m will not produce a more robust algorithm. On the contrary, if the algorithm needs such a fine balance between the two probability parameters this is an indication that the algorithm is unstable. The disruptive parameters that are, however, at our disposal to manipulate are the defining length and order of a schema. These two parameters are defined by the representation format with which we choose to represent the problem domain. Consequently, we should choose a representation that will result in above average fitness schemata being short and of low order.

The issue of disruption is so important that an example is given to stress the effect of disruption and the importance of choosing a 'co-operative' representation. Consider the schema of length 5 $H=\{1***0\}$ over the conventional binary string representation (the most significant bit is on the left and each position thereafter is of decreasing significance), and further assume that this schema has a very high fitness value. The schemata processing would select for reproduction strings belonging to this schema with a high probability due to the relative high fitness value the schema exhibit. Unfortunately, unless such string is crossed with a string that also belongs to the same schema, the resulting offspring will not belong to the said schema and consequently, the propagation of this schema will be substantially disrupted. In contrast to the poorly represented schema H we would choose a representation that will bring the most significant bit from the left closer to the least significant bit, and reduce for schema H the disruption due to crossover.

The Dynamics of Schemata Frequency

The previous section disclosed the secret weapon of GAs – schemata processing – and the fact that search is 'motivated' by schemata fitness although it manipulates and processes strings. Equipped with this knowledge and thus shifting the weight from currently above average strings in the sample to currently above average schemata, it is interesting to study the dynamics of schemata frequency as a function of search progress. By examining which schemata have above average fitness values, one can foresee the population of coming generations due to the fact that these schemata are expected to proliferate. The texture of the future populations can be estimated through schema frequency and fitness. It follows from the expected proliferation of above average schemata at the rate indicated by Eq. (4) that strings of future samples shall instantiate these schemata that proliferate. It further follows that rather than spread arbitrary 2^l schemata as each string does and the whole population of strings do at the initial stages of the search, strings at advanced stages shall have to instantiate the prominent schemata. Finally, because actual strings do not contain the auxiliary symbol '*' and are entirely specific, we can conclude that strings shall reflect the union of schemata. This allows us to stipulate what will be the texture of generations to come.

Let us go back to the black box and illustrate the dynamics of schemata frequency. We can regard schemata as strings and we can draw an auxiliary plot which will plot the domain of schemata (ordered in some fashion), and the range of schema averages $f(H)$ (illustrated in Fig. 18). At the initial stages of the search, the distribution of $f(H)$ is random (however the schemata are arranged). The dotted line marks the average schemata fitness including all schemata (this is also the fitness value of schema $\{*_1, *_2, *_3, \ldots, *_l\}$). As the search progresses, the distribution of schemata frequency becomes more extreme. Schemata that have above average value, the ones that are above the dotted line, grow further and those with below average value diminish. It might be interesting to note that the position of the dotted line is not fixed because it does not represent an independent characteristic of the search space, but rather an expected value estimated by the sample. Because the sample is subject to stochastic sampling, so is the position of the dotted line.

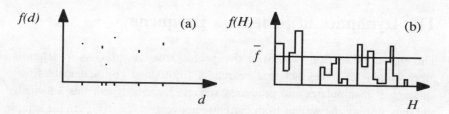

Fig. 18 - *An illustrative sample of search points (a), and the schemata they instantiate (b). The dotted line marks the average fitness of the population.*

One should not seriously consider a practical implementation of such schema counting, however tempting this may be, because a quick arithmetic shows that it is computationally prohibitive. As explained earlier, each string belongs to 2^l schemata that instantiate it. Therefore, it requires $n2^l$ evaluations to calculate the average schema fitness for schemata present in the sample.

No.	Schema
1	11*****************************
2	010****************************
3	***101*************************
4	0001***************************
5	001*****************************

Table 3 *- Five schemata for which the frequency was calculated as a function of the number of trials in the experiments summarized in Figs. 12 through 14.*

Let us choose 5 short, low order, schemata for which we shall calculate the average fitness based on the results obtained from the experiment we conducted with the black box, and let us calculate the frequency and fitness of these schemata as a function of search progress (Table 3 lists the 5 chosen schemata). We have chosen schemata that are not too specific so we expect to have few strings for each of the schemata (at least at the beginning of the experiment). Fig. 19 plots the schemata frequency at 20 trial intervals. Fig. 20 plots the average schemata fitness at the same intervals.

Figs. 19 and 20 show that the reduction in the frequency of schemata 1 and 5 is coupled with their below average fitness (less than 1 in Fig. 20). In both cases, the frequency was eventually reduced to zero. The growth of schemata 2 and 4 can be explained in a similar way. Starting from a fitness value greater than 1, schemata 2 and 4 approach value 1 at advanced stages of the process which indicate that they have become dominant in the population. Schema 3 is somewhat of a 'parasite' schema. Though its fitness value is not high, it is relatively frequent in the sample due to the high frequency of schema 4 which it includes.

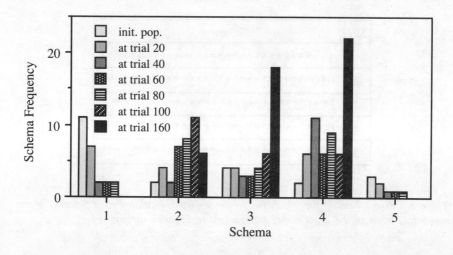

Fig. 19 - *Frequency histogram of the five schemata at different search intervals.*

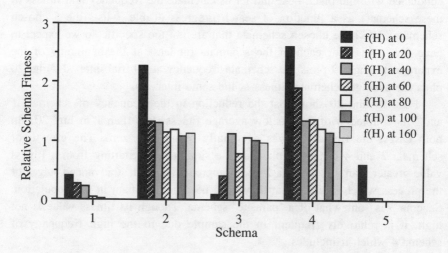

Fig. 20 - *Relative fitness histogram of the above five schemata.*

Why They Are Called Genetic Algorithms

It is clear from the diversity of algorithmic operators employed by users of GAs that there is no precise definition of what constitutes a GA, at least not a definition that is based on its operators. Furthermore, though the diversity of different mechanisms in population genetics is immensely rich, we have no problems in recognizing them as 'genetic'. This suggests that genetic mechanisms in nature have some general underlying concepts which are common to many (all?) of the mechanisms. Therefore, in the introductory part of the book we used the word 'philosophy' rather than the word 'definition' in describing the GAs.

By understanding the schema theorem, the fundamental theorem of GAs, one realizes that it is a procedure which is being emphasized by GAs, not a particular computer code. The most fundamental aspect is the specific structure of information. Nature stores information in simple, even very simple, units and therefore requires a large number of such units to describe complex processes. A GA adopts the benefits resulting from such information structure, and it is that approach to representation of information which makes a GA 'genetic'. Without that particular structure of information, most of the algorithm's operators would lose their meaning and effect.

Schemata processing is another fundamental aspect of GAs that is based on the processing mechanisms of population genetics. The idea is that the fate of individual solutions carries little importance both in nature and for GAs. Whether or not the individual solutions are appropriate, if the schemata that define a particular string have ample presence in the population, then such a string will emerge at some stage, with the exponentially increasing probability of doing so. Clearly, this is also the method by which individual solutions become discarded.

Lastly, we should focus on the role of recombination. It is clear that selection can operate only if it has a variety to select from. The question is: "How does one create variety?" The realization that in nature the gene pole, which is readily available for combinatorial rearrangement through sexual reproduction within a population, has an enormous genetic reservoir, further substantiates the benefits of a chromosome information structure and the use of recombination as means to create variety. This is also realized by GAs as the main process by which variety should be created. Furthermore, mutations (equated here with local search or optimization) have a diminishing importance in providing variety because of the mutual dependence among genes. The mutual dependence among genes results in a 1:1000 ratio working against successful mutations (on average, for each mutation which results in an improved quality, there are about 1000 mutations which reduce the fitness of the organism). Mutations are needed for operators of secondary importance and in situations where the population size is small and thus prone to large statistical noise.

When is the Use of Genetic Algorithms Advocated?

Being new to GAs, researchers and engineers often ask the question: "When should one use a GA instead of other algorithmic strategies?" The answer to this question is not easy because it depends on too many factors which are usually unknown at the time the question is being asked. Nevertheless, we shall attempt to give some directives as to when it is appropriate or promising to use the genetic approach. To do so we have to recapitulate the essence of the genetic approach and suggest some intuitive arguments. Although there is a considerable amount of experience over a diverse variety of domains in applying and using GAs, the theory of GAs cannot offer a conclusive answer.

The most noticeable practical implication of the schema theorem is its implicit parallelism. Because each string evaluation provides information on 2^l schemata it instantiates, the larger the space, the greater the advantage the genetic search has over other procedures. The implicit parallelism suggests a search efficiency in the order of $o(n^3)$ (for a detailed explanation on GAs search efficiency see [Goldberg, 1989b]). This means that for a population size n there are on average n^3 schemata which are being processed efficiently [Goldberg, 1985].

Another important feature of the genetic search is the reliance on global information, *i.e.* the schemata that are being processed and consequently emerge to dominate the population spread over the entire space. Furthermore, the underlying concept of schemata is that schemata are a degenerate description of the entire space. It is the processing of global information that is so lacking in other search procedures, and it is global information that provides search robustness. The global information exchange through GAs suggests that multi-modal, multi-dimensional spaces can be searched and processed with reduced risk of prematurely settling on a local optimal region. In fact, it is only worthwhile to apply GAs to spaces that are complex and nonlinear enough so that the use of targeted algorithms such as the gradient methods is unsatisfactory. The search robustness GAs inherently exhibit suggests that when entrusted with arbitrary problems such as different functions to optimize, GAs have, on average, a good chance of out-performing other algorithms. The fact that ultimately any deterministic function-coding may be reduced to a list of fitness values, makes the underlying string representation a methodology which presents different problems to a GA in the same way.

The Epistasis[2] Dogma

The literature on GAs emphasizes that GAs do not 'see' the problem domain because the latter is obstructed by the representation. Accordingly, the question: "Which problem domains are suitable for a GA search?" should be replaced with the question: "Does the representation (of a problem) promote an efficient GA?" By shifting the question of suitability from the problem domain to the representation, one focuses on the core issue of GA applications, thus asking a question which not only is more consistent with the schema theorem, but is also easier to answer (at least its meaning is clearer).

The schema theorem [Holland, 1975] lists the prerequisite features a schema should exhibit in order to utilize a GA search, namely, a chromosome structure composed of short, low-order building blocks. Although granting the logarithmic emergence of above average fitness schemata, the schema theorem does not provide tools to assess the suitability of a representation. When considering the optimization of real-world problems, one realizes that the representation is the only aspect that can be tailored to the convenience of the programmer. Due to the flexibility and re-programmability of a representation, attention was directed towards establishing criteria for the suitability of a particular representation to a GA, and some analytical tools were developed to assist the calculation of the expected performance [Bethke, 1980; Goldberg, 1988; Goldberg, 1989a]. Although much attention was given to the issue of building blocks, their size and number, the effect which interdependency among parameters has over the GA performance did not receive proper attention [Goldberg, 1988; Grefenstette, 1979]. Only certain degrees of nonlinearity enable a GA search to exhibit a relative efficiency, yet others will diminish this efficiency. Therefore, the amount of interdependency among the representation elements is an important ingredient in the GAs' cookbook, and constitutes an important source of information.

---

[2] Epistasis designates in biology the effect of one gene on the expression of another gene, so that if such an effect exists we say the affected gene is epistatic to the effecting one.

Gene interaction is also a central issue in natural genetics where genes are not only dependent on other genes in order to jointly express phenotypical (resulting organism) characteristics, but also suppress or activate the effect of other genes [Ptashne, 1989]. The term that has become synonymous with almost any type of gene interaction is *epistasis*. Derived from the Greek words *epis* and *stasis* ('stand' and 'behind'), epistasis can be equated with *stoppage* or *masking*. Epistasis is used to describe the situation where one gene pair masks or modifies the expression of another gene pair. When the epistasis of a chromosome is said to be high, it means that many genes are dependent on the presence (or absence) of other genes. It is helpful to remember that GAs, as many natural systems, assume a certain schema structure, a structure where the whole is different from the sum of its parts [Jacobson, 1955; Platt, 1961; Simon, 1962; Tsotsos, 1987; Goldberg, 1989d]. The frequent characteristic of such information systems is that by knowing the value of the parts one cannot accurately calculate their effect together. In the GAs coterie epistasis is used to indicate the extent of nonlinearity and interdependence among the elements composing the representation.

Tracing epistasis is an elusive occupation because the presence of epistatic elements can only be noticed at the phenotypic level away from their scene of interaction. Furthermore, even if epistasis is given, what can one say about a GA search knowing the representation's epistasis? If a representation contains very little or no epistasis, any individual element is not affected by the value of the remaining elements and therefore, optimization becomes a bit-wise maximization. At the other end of the epistatic scale, if a representation is highly epistatic, then too many elements are dependent on other elements and the building blocks become long and of high order. When the epistasis is extremely high, the elements are so dependent on each other that unless a complete set of unique element values is found simultaneously, no substantial fitness improvements can be discovered (such as in the parity problem). Under such extreme circumstances, complexity has exacerbated to the extent that the performance space does not contain significant regularities. This leads to the notion that a representation should be constructed so˙ as to incorporate mild epistasis, neither too high nor too small. In Fig. 21, the three typical search strategies, hill climbing, GA and random search, are plotted on a percentage epistasis scale according to their zones of relative efficiency: low, medium and high epistasis, respectively.

What effect has epistasis on relative efficiency? A representation with small epistasis means that co-adaptation is not prominent and therefore, a hill-

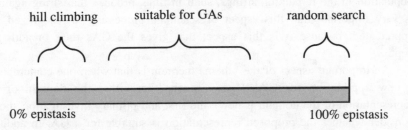

Fig. 21 - *The epistasis scale, and the epistasis regions suitable for hill climbing, GAs and random search.*

climbing algorithm is more likely to be more efficient. A representation with high epistasis implies that co-adaptation is very strong and therefore, the efficiency of a GA will decrease significantly. A representation with mild epistasis is suitable for a GA (Fig. 21). It is now possible to appreciate that there are additional factors which affect the expected performance of a GA besides the nature of the building blocks. If the epistasis could be calculated for a given representation, then it would offer an important yardstick of its suitability to a GA.

Summary

Now that we are familiar with the schema structure and the schema theorem it is easier to understand the motivation in representing a domain with an auxiliary parametric space. As a matter of fact, we increase the efficiency of the search and its processing powers when we choose a representation which is as detailed as possible, and as long as the representation is still natural to the domain investigated. We have also learned that although the algorithm only manipulates individual strings, it is in fact the 'background' growth of schemata that determines where search efforts should, and will, be invested. Because schemata are dependent on a relatively large number of strings, they are much less vulnerable to noise and stochastic errors. That is the reason why we can choose to be liberal and allow the possible disappearance of very good strings through the random selection for replacement. If the strings are in fact globally good, then the schemata they instantiate which are left in the population in the remaining strings, shall in time, produce that string again. It is very important that this aspect of schemata processing is recognized and appreciated, because it is this aspect that gives the GAs their breadth and depth.

An important aspect of the schema theorem is that when one contemplates a new application for GAs ('new' means that either the domain or the representation of the domain is new), the first and primary investigation should consider whether the proposed representation is suitable for a GA. In addition to the directives given by the schema theorem on the nature of the building blocks (that they should be short and of low order), the epistasis analysis gives some notion of the amount of nonlinearity a suitable representation should contain. With the two perspectives on analyzing GAs efficiency described in this chapter, we have improved our intuition regarding the kind of representation spaces that lend themselves to an efficient search. These are complex, large spaces that have some structure in them.

Observing the complexity found in nature, and the success with which population genetics directs the evolution of species to be better fitted to their environmental niche, one can draw an analogy with many optimizing problems. For most complex applications, the search for co-adapted control parameters is a very demanding pursuit. The analogy with natural adaptive mechanisms is the backbone of a new family of probabilistic heuristic algorithms called Genetic Algorithms – efficient search-for-the-optimum procedures that use simplified models of sexual reproduction and selection to produce new solutions. From the mechanism of GAs and from the schema theorem it is indicated that GAs will prove to be efficient when nonlinearity is present, but not too strongly. This notion was further substantiated by the analysis of epistasis embedded in a representation.

Chapter 3
A BRIEF HISTORY OF
GENETIC ALGORITHMS

Neither the paucity of researchers in GAs, nor the shortage of GA applications, has caused this chapter to be brief. On the contrary, the diversity of applications incorporating GAs is one of the more impressive aspects of the utility of GAs. A fair account of the ample work and of the advances both in the theory and the application would simply take too long, and is beyond the scope of this book. A detailed review of the history of GAs can be found in Goldberg's book [Goldberg, 1989b]. Inquisitive readers shall also find abundant material in the international conference proceedings on GAs, and other publications. We shall, therefore, limit our review to important milestones in the general history of GAs, and to a more detailed review of that work which relates to the subject of this book – a GA model for robot trajectory planning.

The History of Genetic Algorithms

Professor John H. Holland, in his book *Adaptation in Natural and Artificial Systems*, sets the framework for this special approach to search and adaptation. In a way, the publication of the book marks the birth date of GAs, or at least when they 'went public'. Holland's book is a projection of separate ideas and realizations on which he worked for many years preceding the book's publication. Starting with broad outlines of adaptive systems during the early 1960s, Holland then proceeded to analyze the role of recombination in adaptation [Holland, 1971], and the role of schema processing [Holland, 1973]. Then, in 1975, the time was ripe to join all the separate links into one chain, and the '75 book does exactly that. Notwithstanding contributions made by other researchers (mainly students at the University of Michigan, the 'Mecca' of GAs for a long time) which bear the theme of genetic algorithms [Bagley, 1967; Cavicchio, 1970; Rosenberg, 1967], GAs are rightly associated with Holland. Until the early 1980s, the research in GAs was mainly theoretical, with few real applications. This period is marked by ample work with fixed length binary representation in the domain of function optimization [De Jong, 1975; Hollstien, 1971]. Hollstien's work provides a careful and detailed analysis of the effect different selection and mating strategies have on the performance of a GA. De Jong's work attempted to capture the features of the adaptive mechanisms in the family of genetic algorithms that constitute a robust search procedure. For this purpose, De Jong devised a suite of functions and focused on applying a rigorous computational analysis to that function suite. The functions were carefully chosen to exhibit the classical function features search algorithms deal with, namely: continuity, modality, dimensionality, stochasity and convexity. De Jong's work in formalizing the framework for simple GAs directed many subsequent developments in GAs to follow his footsteps.

The research into the workings of GAs and the theory of their mechanisms continued. From the early 1980s the GA community has experienced an abundance of GA applications which spread across a large range of disciplines. Each and every additional application gave a new perspective to the theory. Furthermore, in the process of improving performance as much as possible via tuning and specializing the GA operators, new and important findings regarding the generality, robustness, and applicability of GAs became available. In the engineering field, Goldberg's work on steady-state and transient optimization of a gas pipeline using GAs is a classic example [Goldberg, 1983]. There are other notable works in engineering [Davis and Coombs, 1987; Fourman, 1985; Glover, 1987; Goldberg and Samtani, 1986], pattern recognition [Englander, 1985; Grefenstette and Fitzpatrick, 1985; Stadnyk, 1987; Wilson, 1985], neural networks implementations [Ackley, 1985; Cohoon, *et al.*, 1987; Dolan and Dyer, 1987; Jog and Van Gucht, 1987; Suh and Gucht, 1987; Tanese, 1987].

Previous Work Relating to the Main Issues Raised in this Book

The most central issue in the application of GAs to robot trajectory planning is the requirement to use flexible representations and the order quality of trajectory generation. In the context of GA applications, the work of Smith [Smith, 1980; Smith, 1984] on dynamic representations is most similar to the type of representations proposed here. Smith's Learning System (LS-1), which is based on a Classifier System type GA [Holland and Reitman, 1978], constructs a complete production program from a variable number of production rules. In LS-1, one production program is one chromosome (one string). This approach to production systems (PS) differs from the earlier model of PS in that the genotype is held together as a whole. The differences between the two approaches run very deep both in theory and in practice, but mainly affects co-adaptation among representation elements and issues of credit assignment.

LS-1 proposed to overcome the 'unfixedness' of string lengths by incorporating multi-level reproduction operators. Reproduction in the presence of strings of varying length necessitates extensive modifications of all reproduction operators, but mainly to the crossover (see illustrative example of Chapter 1 or a detailed discussion in Chapter 6). LS-1 differs from the model which will be proposed here in that it assumes order independence among rules.

Another interesting model that used dynamic string lengths was presented by Shaefer in his ARGOT strategy [Shaefer, 1987]. In this work, an adaptive intermediate mapping is used to translate the chromosome into a solution in the phenotype space. Each parameter has an internal representation composed of a varying number of binary bits. The ARGOT system is capable of dynamically altering the resolution, the scale and centre of each individual parameter range. The dynamic representation is driven by a set of new (to the standard GA optimizer) operators whose effect on the representation is Lamarckian in nature. The ARGOT strategy can neither be applied to order dependent representations, nor to representations composed of a varying number of parameters. Other models similar to the ARGOT strategy were developed to determine consumer choice [Greene and Smith, 1987], and GENES, a model of GAs that creates and uses lists of tree-structured production rules [Bickel and Bickel, 1987] used variable string length, *i.e.* considered flexible quantities of information.

Goldberg recently introduced a somewhat different type of GA, a messy GA (mGA) [Goldberg, *et al.*, 1989e]. Messy GAs process variable-length strings that may be either under- or over-specified with respect to the problem being solved, using less rigid operators than is the normal practice. Goldberg realized that an important factor in nature's robust adaption is flexibility of information representation. This observation is translated into practice through the introduction of messy codings, messy operators, and cursory capability for gene expression. A fundamental feature that enables mGA to emulate "...nature's climb out of the primordium...", as Goldberg puts it, is the capability to construct and test simple organisms and then use these tightly linked combinations of features to construct more sophisticated structures. Messy GAs represent a fundamental change in the methodology and are only in the first stages of development. Nonetheless, first results are promising, and point to an avenue of research in variable representation that may be useful in increasingly complex problem domains.

The Lamarckian theory – the evolutionary theory that postulates the inheritance of acquired characteristics – lost its followers when modern research did not support the principal arguments of the theory. However, various mechanisms were suggested where such directed learning can enhance general machine learning. The most relevant models are ARGOT [Shaefer, 1987; Shaefer, 1988], VEGA [Schaffer, 1984; Schaffer, 1985a], and the adaptive crossover distribution GA [Schaffer, 1987]. ARGOT exhibits clear Lamarckian qualities whereby performance changes the internal representation which alters the genotype structure passed to successive generations. In the multi-objective evaluation technique used in VEGA (LS-2), Lamarckian characteristics were exhibited in the way global performance is decomposed into sub-objective components to enable a better assessment of the genotype building blocks. Schaffer's work on an adaptive crossover distribution mechanism is most similar to the model used in this work. The crossover probability distribution GA was implemented in an indirect way by storing a cross site topology on a second genotype attached to the main genotype. The distribution of crossover probabilities along the string were determined implicitly. The main differences between the above works and the Lamarckian operators used here, are that the information extracted from present performance is directly associated with reproduction operators, and is applied explicitly.

The use and benefits of incorporating redundant parameters in a string representation was addressed in the work of several researchers, but mainly under various models of diploid representations [Goldberg and Smith, 1987; Holland, 1975; Hollstien, 1971]. In these chromosome representations, the information is stored twice, although expressed only once. The reasons for adopting diploid representations are rooted in the idea that a robust GA, especially in dynamic environments, should incorporate a long term memory, longer than one current policy. In this work, the aim of incorporating redundant parameters is solely to increase representation robustness in the face of inexperienced programmers. The main objective of the diploid representations is to be able to operate within a dynamic performance environment.

This book addresses the issue of robot trajectory planning via GAs. Thus far we have looked into GAs *per se*. The reason for introducing GAs to the robotic environment was that there was a belief that GAs might prove to be a rewarding strategy to employ in solving problems in that domain. This belief was also realized by another application of GAs to the robotic world, seen in the work of Khoogar [Khoogar, 1987; Parker, *et al.*, 1989]. In his work, Khoogar has applied a GA to the problem of inverse kinematics. The problem of inverse kinematics will be explained in greater detail in Chapter 4, but we may note that in general the problem of robot arm inverse kinematics has no unique solution. Khoogar used a GA to solve the problem of inverse kinematics for a three-link spatial robot arm and then extended his model for redundant structures. There is little in common between the work of Khoogar and the work presented here although both address the problem of optimizing the positioning of a robot arm. The difference between the two works can be best described as the difference between statics and dynamics. In trajectory planning, not only does the dynamic behaviour of the robot have to be considered, but so do the aspects of order dependent sequence forming and performance in the physical space.

PART II

A GENETIC ALGORITHM FOR OPTIMIZING ROBOT TRAJECTORIES

Chapter 4

THE ROBOT ENVIRONMENT

Robotic research and advanced automation have thrived during the last two decades, introducing a totally new and revolutionary approach to manufacturing. The main impetus has been the realization that robotic technology provides a flexible mechanical structure that can perform a wide variety of tasks in a re-programmable fashion (the industrial equivalent of a 'Jack of all trades'). However, to produce such a flexible yet economical machine, the system's components: mechanical structure, motion actuators, control, and programming language, must be relatively sophisticated. This results in a system which is highly complex, nonlinear, tightly coupled, and multi-dimensional; a system that for any non-trivial application requires extensive programming and optimization.

This chapter aims at exploring the underlying principles of robotics and discussing its major characteristic problems. It is hardly necessary to show why both such systems and the way they are presently being utilized can greatly benefit from the application of optimizing techniques. It shall be shown here how this can be achieved by means of GAs and the role they may play in improving the performance of robot systems. It shall further be suggested that the characteristics of robotic systems are an archetype for a broader family of processes in the manufacturing environment. The benefits GAs may introduce to robotics may therefore be extended to other similar processes of a similar type.

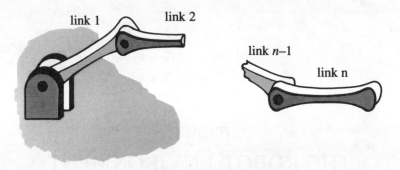

Fig. 22 - An open n-link kinematic chain.

What Makes Robots Tick[3]

A good starting point is a description of the system in terms of its major components and its operation methodologies. Mechanically, a robot arm is an open kinematic chain composed of relatively stiff links and a joint between adjacent links (Fig. 22). Since a robot arm performs a task through the motion of its end-effector attached to the last link; the last link is the primary component of the whole structure (Fig. 4). Each joint represents one degree of freedom (standard systems have six degrees of freedom to provide full spatial flexibility). The joints, either revolute or linear, are individually held in position by actuators such as an electric motor, or a hydraulic cylinder. The actuators drive all links independently. Obviously, the individual link motions must be synchronized so that the whole structure will move in a controlled fashion. The actuators are connected to a processor which is called the motion controller. This controller communicates with the actuators in duplex, sending displacement commands and receiving feedback on their progress. Controlling multi-axis motion is a very demanding task and poor performance, where it exists, is soon manifested when the system is set in motion. Too often, a robot is viewed as a mechanical structure that moves in a precise, predetermined

---

[3] The title for this section was adopted from the title of a paper by the late Professor J. Hatvany of the Hungarian Academy of Sciences [Hatvany, 1987].

way. In practice, the dynamic behaviour of the arm is complex, being influenced by many internal and external factors. For example, most robot structures are mechanically coupled, and a movement of one link induces reaction forces on the adjacent links through the connecting joints. The induced forces result in a complex dynamic response whose effect is very difficult to predict.

The design of robot systems makes them relatively easy to program to do something close to what is desired, yet extremely difficult to make them do exactly what is desired. This is due to the unpredictable factors mentioned above. The discrepancy between the desired and the achieved is mainly the result of the division of labour between the human planner and the computer. This division of labour is such that the human planner gives only a general description of the desired path, leaving the computer to work out the detailed trajectory of motion. Trajectory generation is divided into two distinct phases:

(1) The human interface by which the programmer specifies a path.
(2) The machine dependent automatic translation of these path specifications into a detailed motion plan as a function of space and time.

It is easy to fall into the above trap because we wish to specify paths as simply as possible and let the computer shoulder the major load in computing the trajectory. We shall discuss these two programming phases, and the reason why the computer cannot perform by itself what we expect. However, first we must acquire a better understanding of what a robot system is.

Due to the all too understandable preference that the human interface shall be as simple and intuitive as possible, robot paths are specified as a sequence of positions the end-effector is required to visit. Because the motion of the robot arm consists of the motion of the individual links, the motion controller analyzes the specified end-effector trajectory in terms of motion information for each of the links (the spatial position of the links in a given stationary arm position is called an *arm-configuration* because it configures a unique geometrical arm structure). Commanding the arm to move from one arm-configuration to another is in fact issuing a displacement vector that specifies the net joint angle change that each link is required to undertake. The sum total of the link movements should take the end-effector from its current position to the target position.

Computing the position of the end-effector according to the position of the links is called *direct kinematics*, and clearly involves a trivial computation. On the other hand, computing the position of the set of links which results in a specified end-effector position is not computationally trivial. Calculating the arm-configuration which specifies a given end-effector position is called *inverse kinematics*. Inverse and direct kinematics emphasize the transition from the path specification determined by the human programmer and the work the computer must perform to translate this information (inverse kinematics) into detailed trajectory instructions (direct kinematics). Another major problem in trajectory generation of continuous motion is that most robot programming is done in off-line and discrete stages. The advantage of trajectory specifications being stored in memory, and which can thus be re-programmed and edited, is balanced against all the common problems of creating continuous motion from discrete information.

We have briefly discussed trajectory generation, but have already identified one issue as critical. Whatever formula the computer is assigned to follow when generating the complete trajectory from the partial specification given by the programmer, the produced trajectory will predominantly be dependent on the programmer's specifications. According to this matrix, and while a human programmer is providing the specifications, the only aspect which can be optimized in the trajectory generation process is the computer algorithm. Indeed, this is a well-researched subject (interested readers are encouraged to refer to some good reviews given in [Asada and Slotine, 1986; Craig, 1986; Paul and Zong, 1984]). Although much progress can be made by improving the computing schemes that convert the sequence of specified end-effector positions into a defined trajectory, it is not the type of optimization with which the present study is concerned. The fact that the computer algorithms are invisible to the programmer is recognized in this work as the major reason robot systems perform poorly. The 'bottle-neck' in optimizing robot trajectories lies in the programmer and the original path specification. We shall address this issue further at the end of this chapter and in the following chapters.

After the programming phase has been completed (a sequence of arm-configurations has been recorded), the program is tested to enable the programmer to check the resulting performance. This 'checking' phase is necessary because the programming is a discrete process and done off-line, thus avoiding the powerful dynamic effects to which the system is subject under normal working conditions (the current state of robot programming is a good example of the phrase 'the proof of the pudding is in the eating'). Programmers cannot accurately foresee the effect of their programmed application. There are additional factors that contribute to poor robotic performance. The following section will discuss two aspects of the human interface which are considered to be a major source of poor performance.

The Human Interface

The operator, when entrusted with a path the robot is required to follow (the desired path), has to divide the continuous path into discrete arm-configurations so that the motion between them will best approach the desired path. The number of different possible trajectories that will result in an executable trajectory similar to the desired path is very large. No two programmers will specify the same trajectory. The trajectories they specify will be different either in some or all of the arm-configurations. The trajectories may also differ in the number of arm-configurations that compose the trajectory. There is a reason for the large number of possible trajectories. The reason is that specifying a path is usually an under-determined problem (the size of the trajectory space at the stage of specifying it will be analyzed in the next chapter), and there are no algorithms to specify a path.

Clearly, different trajectories may have different performance qualities (whatever 'quality' means in a particular application). The programmer wants to find the optimum or at least the near optimum trajectory from the many possible trajectories. Yet he is unable, as he has no means to predict which path specifications will result in an optimum trajectory once they have been processed by the system, unless of course the different trajectories are tested. The excessively large size of the trajectory space leads to *ad hoc* programming which ignores substantial sections of the trajectory space, sections which may very well contain near optimum trajectories. The experienced robot programmer will readily confirm that the various 'near' optimum trajectories programmers create are usually quite distant from the really good performances that can be obtained.

Let us consider some of the aspects of trajectory specification which make prediction of the resulting trajectory difficult. A fundamental interface problem is the difference between the trajectory specification taught off-line, and the path resulting from the continuous motion. At the specification phase the trajectory is constructed from discrete arm-configurations where the dynamic properties are not present. In fact, many additional factors which influence the trajectory-to-be are not present at this stage and therefore cannot be considered, nor their effect counterbalanced (by way of example see the obscuring effect of the interfacing language as demonstrated in Fig. 23). When the programmer constructs the trajectory off-line, he can only speculate on the outcome of the programmed trajectory because the dynamic behaviour, coupling and inertia, are so extremely complex that they cannot be visualized by the human mind with sufficient accuracy.

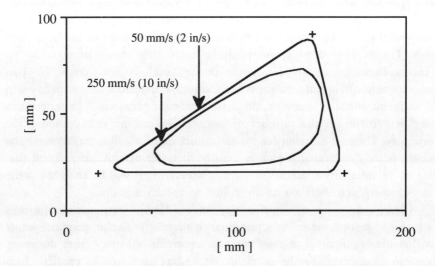

Fig. 23 *- Two trajectories attempting to follow a triangular path marked by the three crosses. The reason for the unexpected deviations from the desired path lies in the system which gives precedent to speed. Thus, when speed is increased, position accuracy is compromised.*

Another major interfacing problem consists of the difficulties in human perception of different reference frames. One of the frequently used programming methods is the teach-in off-line programming. The programmer leads the robot arm manually to the desired locations and manipulates the arm to obtain the desired position and orientation of the end-effector. For a human, it is fairly easy to control spatial positioning such as the Cartesian co-ordinates X, Y and Z. On the other hand, spatial orientation perception of humans is poor, especially with non-Cartesian systems such as cylindrical or spherical systems. Much time is spent and unnecessary constraints are introduced by the programmer who sees the world through the Cartesian frame of reference, but has to specify revolute co-ordinates.

So far we have limited our discussion to the determination of systems. However, many applications and systems operate under redundant conditions which make the specification of a path so much harder to optimize. Applications that do not require the full spatial flexibility of the arm, leave the programmer with yet another problem – the decision as to which links are redundant. Often, the under-determined problem is simplified and solved by the artificial imposition of additional performance specifications which were not required by the application and consequently unnecessarily restrict the system. Thus no use is made of the flexibility robot designers were so eager to provide and the utilization of the system is reduced unnecessarily.

The under-determined properties of the robot arm, the manipulation of non-Cartesian axes, the dynamic performance discrepancies and dynamic coupling effects, make trajectory specification an art. An art that, alas, produces in too many cases a poor performance. Traditional optimizing techniques and control methodologies such as model reference control, self-tuning regulators, etc., depend entirely on explicit control laws, control laws that require an understanding the programmer usually does not have, nor has the time to acquire [Mars, 1989]. The remainder of this book will propose a GA which optimizes robot trajectories by optimizing the trajectory specification. Before going into the robotic side of trajectory optimization we shall look into the general aspects of robotics that have affinity with the salient points of GAs, and into the fundamental problems in applying a GA to the trajectory optimization domain.

Genetic Algorithms in Robotics

In the preceding sections, robotic systems were described as extremely nonlinear and complex systems for which no straightforward algorithms exist to predict their dynamic behaviour. In Part I of this book, GAs were presented as adaptive mechanisms that are successful in handling certain complex and large domains. Are the problems of robot systems the type of problems GAs are relatively efficient in solving? The present section will address this possibility, and suggest several areas where GAs can enhance robot performance. The discussion here will be brief in order to avoid an unnecessary repetition of issues which are dealt with at length by the remaining chapters of the book.

There are three areas to which optimization can be applied in order to enhance robot performance:

(1) The original system design (mechanical, electrical, control, etc.).
(2) The programmable hardware parameters for individual systems.
(3) Trajectories and motion programs for individual applications.

Robot systems are recognized to be systems that can benefit greatly if the individual sub-systems, such as the mechanical structure, control circuitry, etc., are optimized [Lenarcic, *et al.*, 1987]. At the design phase robot systems are highly under-determined systems. The designer has to establish a large number of arbitrary design specifications in order to reduce the number of degrees of freedom to a manageable size. The large number of redundant design specifications often hinder flexibility and performance. GAs can be applied to optimize the design specifications. A straightforward example of such early design optimization can be appreciated when considering the geometry of the mechanical structure. The dimensions of the arm links, and the working sectors of the joints determine the characteristics of the *working envelope*, the area defined by all the positions which the end-effector can reach. The parameters of the working envelope have a clear impact on the utilization of a system. In addition to the working envelope, many other performance features depend on the mechanical design. GAs could optimize some of the aspects of the mechanical design without the necessity of physically building many robots. A recent feasibility study of optimizing the mechanical design of some components of robot systems with a GA shows promising results.

Once a robot system is built, GAs can be used to optimize the programmable control parameters such as gain, threshold, etc. They can also improve the response and performance of the system in a specific working environment, or in specific operating conditions. This is a more temporary optimization since it is carried out for a specific task, and only manipulates the system's input signals while the hardware of the system is left unchanged. The third area in which GAs can improve robot performance is in optimizing trajectory specification. Trajectory specification has been left to the human operator in most robot applications and, as discussed in previous sections of this chapter, the human interface presents a major problematic aspect of robot applications.

The areas of application can be viewed as layers of hierarchical optimization. The first is a low level optimization of the system's basic design, aimed at producing an optimum design for a variety of tasks and working conditions. The second is an intermediate level, aimed at optimizing a system for the working environment of the customer. High level optimization aims at making the most from what the existing system has to offer for each individual task. One may argue that it is not economical to employ optimizing algorithms for each new application. This would be true only if *ad hoc* methods produced a satisfactory performance, an all too infrequent state of affairs in the robotic environment.

Summary

This chapter introduced robotic systems, their special features, methods of usage, and type of applications. Robot systems were presented as an archetype for manufacturing machines and order dependent processes. After identifying the problematic features of robotic systems, the areas where GAs can improve robot performance were described. It was argued that the path specification domain is an area that can benefit from algorithmic strategies for its optimization and that this area did not receive much attention from the robotic community. In the following chapters we shall attempt to offer an algorithm, a GA, that is capable of specifying near optimum trajectories.

Chapter 5
A SIMULATED ROBOT SYSTEM

We shall develop a GA and experiment with it to achieve a trajectory generation on a simulated robot system. Because of safety considerations, it is advantageous to use the controlled environment of a simulated robot system rather than have a several ton robot train itself to follow a specified trajectory by waving its arm in a random fashion through a number of trials, however few. This is an important pragmatic issue which should be disposed of at this early stage. The goal of this work is to investigate the applicability of the concepts of genetic adaptive search to the complex domain of trajectory generation. As such, the direct applicability of the illustrative system is largely irrelevant. Nevertheless, the use of a simulated model rather than a real robot should not be regarded as the means by which we over-simplify the problem or skillfully avoid the main issues which were pointed out in the previous chapter. Our interest in addressing complex problems which do not have a straightforward algorithmic solution is restated here. It is not claimed here that the trajectories which will be planned for the simulated robot shall be optimal or even applicable to any real system, and it is not our intention to optimize any specific trajectory. What we are interested in is a realistic environment which captures the essence of robot trajectory generation so that we can test our optimizing algorithm in as realistic and as universal as possible a system.

To emphasize this aspect the simulated model does not represent a specific or even a realistic system. While accentuating all basic dynamic effects which are encountered in real systems, the simulation uses only elementary control commands, which in practice would make the system cumbersome to operate. The use of such a 'primitive' control model turned out to be serendipitous because it broadened the plausible utilization of the developed GA to most robot systems and constituted a general manufacturing process archetype.

The simulation package was developed by Dr Leon Zlajpah of Josef Stefan Institute, Yugoslavia as a general purpose simulation to investigate the dynamics of a robot with 3 degrees of freedom. At present, the package is tailored to suit the specific research requirements, namely, to offer a complex, nonlinear and redundant system. It is important to note that the simulated robot model is a redundant structure, incorporating 3 degrees of freedom in a plane, the vertical plane, to include the effects of gravity. It is interesting to speculate how successful a GA might be in handling a redundant structure. Because today's robotic industry strives to integrate multi-axis robot systems, adopting a redundant model was particularly relevant to this industrial discipline.

The detailed mathematical description of robotic motion is a fascinating subject. However, for the purpose of investigating general trajectory generation and not to reduce the universality of the model, the robot system must be treated as our black box from Chapter 1. Any excess information about the specific workings of the simulated model might negatively influence the development of a general GA by specializing the algorithm to this particular model and application. Whether it is a true robot performance, or a badly simulated system, a properly constructed GA would optimize the performance to the extent of available information. This robustness of GAs is what this book intends to study, and it must not be confused with the performance of the simulated robot model.

The Robot Simulation Package

The *working envelope* is the geometric territory reachable by the end-effector (see an illustrative working envelope in Fig. 24). In the simulated model used in this work, the three links are mechanically restricted to movements in the

Fig. 24 - *The working envelope of a 3-link planner robot (shaded background). The movement of the links is limited to the sectors between the lines extending from the centre of the respective joints.*

vertical plane. Hence, for all end-effector positions strictly within the working envelope, one link is redundant. All links are 1 m in length, 20 kg in weight, and with a large moment of inertia. These design specifications produce a very sluggish arm (Fig. 4). Each link is free to rotate 360° about its joint, resulting in a working envelope defined by a circle with a 3 m radius around the arm origin.

An arm-configuration $\mathcal{A} = (\mathcal{A}_1, \mathcal{A}_2, \mathcal{A}_3)$, is a unique arm structure defined by the three link positions incorporated in the structure (Fig. 25). Given the triplet of link positions, the end-effector's position is uniquely determined.

Without going into the details of the dynamic equations used in the simulation (more details about the working of the dynamic simulation are presented in Appendix A), it will be noted that all dynamic effects, besides link elasticity, are considered: gravity, centrifugal and coriolis forces, backlash in the gears, dynamic and static friction of the DC motors, and some controller side-effects.

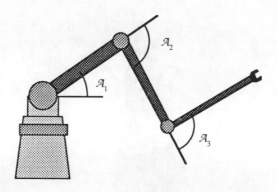

Fig. 25 - *Link position in Euler's angle notation.*

A *displacement vector* $\Delta\mathcal{A} = (\Delta\mathcal{A}_1, \Delta\mathcal{A}_2, \Delta\mathcal{A}_3) = (\mathcal{A}_1^T - \mathcal{A}_1^C, \mathcal{A}_2^T - \mathcal{A}_2^C, \mathcal{A}_3^T - \mathcal{A}_3^C)$, is a set of three angle changes $\Delta\mathcal{A}_1$, $\Delta\mathcal{A}_2$, and $\Delta\mathcal{A}_3$, corresponding to links 1, 2 and 3 respectively (the superscripts T and C denote the target and current positions respectively). A displacement vector involves two arm-configurations, the current one and the target arm-configuration to be reached.

When a displacement vector is executed, the three link positions of the current arm-configuration are subtracted from the corresponding angles of the target arm-configuration. The motion vector followed by the simulation is comprised of the change of position of each link. The end-effector motion is the superposition of that motion vector. Each link moves asynchronously with respect to the other links, endeavouring to reach its net angle change as quickly as the kinematic speed and acceleration profiles permit. This asynchronous motion adds to the complexity and nonlinearity of the simulated robot model (Fig. 26). Although the programmer of a real robot application has sophisticated motion commands at his disposal, the simulation uses the very basic motion commands for the reasons set out above.

The input into the simulation is a string of triplets, triplets of link positions defining arm-configurations. The simulation computes the displacement vectors, and executes them sequentially (Fig. 27). Once a target arm-configuration is achieved, the next target arm-configuration is processed.

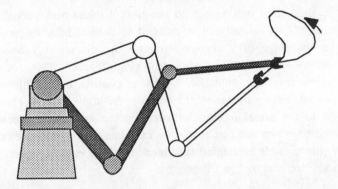

Fig. 26 - *Although the end-effector positions of the two arm-configurations are not spaced too far apart, the resulting end-effector path is badly distorted due to the vigorous link motion.*

The transition from executing one displacement vector to another occurs continuously while the arm is in motion. Because the system has a finite positioning accuracy (≈15 cm), once the end-effector is within 15 cm radius

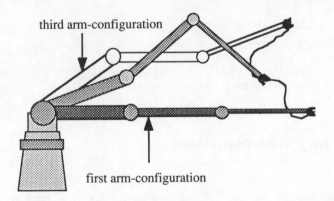

Fig. 27 - *A 3-trajectory defined by three arm-configurations and the resulting end-effector path.*

from the designated position, the controller 'assumes' that the end-effector has reached the target. Therefore, once the end-effector is within the accuracy radius, the simulation will switch to the next displacement vector until no vectors remain. Consequently, it is difficult to predict both the end-effector position at the stage of a changeover, and the discrepancy between the programmed trajectory and the desired path (Fig. 28).

The output from the simulation is a list of pentads, link positions and the end-effector's x, z co-ordinates recorded during the arm's motion. These values correspond to the actual motion followed by the arm while executing the programmed trajectory and can be used to calculate the performance of a given trajectory (the possible evaluation functions and the specific function we used in this model are analyzed in Chapter 6).

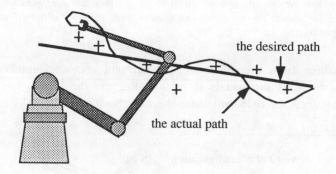

Fig. 28 - *The desired path (straight line), the programmed trajectory (the crosses mark the position of the end-effector), and the actual path (winding line).*

Trajectory Representation

The programming language for the simulated robot (like most robot path-planning languages) requires the discrete end-effector positions to be expressed in terms of corresponding arm-configurations. Thereafter, trajectories are formed by joining several arm-configurations. We define an n-trajectory as a trajectory consisting of n arm-configurations ($n-1$ displacement vectors). It is important to note that this definition is based solely on the number of arm-configurations a trajectory contains. The n-trajectory has the elaborated form

$$\{(\mathcal{A}_1,\ \mathcal{A}_2,\ \mathcal{A}_3)_1\ (\mathcal{A}_1,\ \mathcal{A}_2,\ \mathcal{A}_3)_2\ ...\ (\mathcal{A}_1,\ \mathcal{A}_2,\ \mathcal{A}_3)_i\ \cdots\ (\mathcal{A}_1,\ \mathcal{A}_2,\ \mathcal{A}_3)_n\}\ ,$$

where $i = 1, 2, ..., n$ designates the order of execution according to the ascending value.

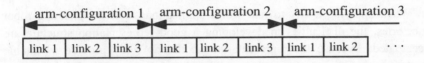

Fig. 29 - *An illustrative sequence of arm-configurations for the 3-link structure.*

We can also view each triplet of joint angles as a characteristic unit of information or the 'gene' of a trajectory. One must admit that viewing an arm-configuration as a gene has some structural beauty because a building block structure is obtained with which we can compose any n-trajectory (see Fig. 29 for illustration). Similarly, we can decompose any n-trajectory into n structurally similar units. A building block structure is a desirable property in any knowledge representation application, and is especially important in GAs (for schema and schemata hyperplane analysis).

We shall adopt for the moment this 'natural' structure for representing trajectories, the same format that is given to the simulation, and investigate whether it is amenable to a GA. It will be advantageous to use the natural representation of trajectories which do not require complicated mapping functions to convert the auxiliary representation into a trajectory format. In his discussion about Broadcast Language, Holland contends that a representation should have the following characteristics:

(1) Position freeness.
(2) Small number of values at each position.
(3) No long-term limitations on the adaptive plan.

The rationale for the first two requirements is clear, and has been demonstrated both analytically and experimentally [Bethke, 1980; Goldberg, 1989b; Holland, 1975]. The third requirement, however, is somewhat elusive. Holland explains this issue in a discussion on fixed representations :

"To this point the major limitation of genetic plans has been their dependence upon the fixed representation of the structures (the set of control signals)...that a fixed representation has limitations is clear from the fact that only a limited number of subsets of... (the control signals)... can be represented or defined in terms of schemata based on that representation." ([Holland, 1975, p. 141]).

When applying a GA to real world applications, such as planning robot trajectories, the difficulties in developing a good representation structure are accentuated. Because the representation should not only provide a suitable structure for a GA, but also conform to the physical properties of the task, it is important to understand these properties and their meaning in terms of GA processing. A partial insight is acquired when we analyze the essential elements for trajectory optimization and see whether it is possible to express these elements in a standard GA representation. The following four elements of trajectory representation should be stressed:

(1) A trajectory should be represented by strings of varying lengths to allow for a varying number of motion vectors to describe a trajectory. This varying length feature of trajectories cannot be described by a fixed length string representation used for standard GAs. Furthermore, the recombination operator assumes homology between any two strings which no longer is true for the natural representation of trajectories.
(2) The order dependency of motion vectors implies that the relative position of a motion vector in a trajectory has a diminished significance. This also means that the simple recombination operator cannot be applied.
(3) The trajectory basic unit of information, the arm-configuration which repeats n times, has a large range of values. This situation was indicated to be counter-productive for GAs.
(4) Due to the natural representation of arm-configurations, the representation contains repetitive sub-structures which must not be separated. This string structure is not accommodated by standard representations.

Because it is advantageous to use the natural representation of real world domains, we shall set this as a temporary goal for this work. The following chapter suggests a model with which the problems introduced by the natural representation of trajectories can be solved, but first we have to analyze several other important aspects of the trajectory space.

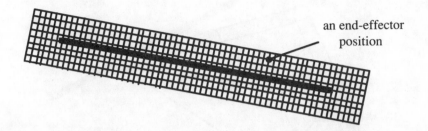

an end-effector
position

Fig. 30 - *The end-effector space in the vicinity of a straight-line path. Each node on the grid marks an end-effector position.*

The Size of the Trajectory Space

The size of the trajectory space is an important element. If there are only a handful of trajectories that can plausibly produce the desired path, then one only has to evaluate them all, and choose the best one. The present section will investigate the size of the trajectory space for the simulated robot model and will show that even for moderate path lengths, the trajectory space is excessively large. We shall define three simple terms which will allow us to estimate the size of a trajectory space for any given path. These definitions can easily be extended to any robot system.

(1) *Arm-configuration space*, A_s, is the set of all mechanically attainable arm-configurations which result in a given end-effector position.
(2) *End-effector space*, E_s, is the set of all end-effector positions within a given boundary around the desired path (see Fig. 30 for an illustration).
(3) *Trajectory space*, T_s, is the set of all n-trajectories having n end-effector positions selected from E_s (we shall limit n to some maximum value n_{max}).

The size of the trajectory space is therefore,

$$T_S = A_S \sum_{n=2}^{n_{max}} \binom{E_S}{n} .$$

(6)

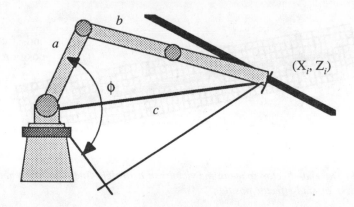

Fig. 31 - The angle range for link 1 at a given end-effector position $E_i = (X_i, Z_i)$.

The trajectory space is dependent on the following four factors:

(1) The length of the desired path (necessary to determine E_s).
(2) The arm's maximum steady-state error (necessary to determine E_s).
(3) The smallest mechanical displacement the robot arm can perform (necessary to determine E_s).
(4) The number of alternative arm-configurations for each end-effector position (necessary to determine A_s).

Let us first estimate A_s by fixing the position of the end-effector and multiplying the binomial coefficients for each of the links:

$$A_s = (\text{positions for link 1})*(\text{positions for link 2})*(1).$$

The positions open for link 1 are the angle range open for the link divided by the angular accuracy that can be attained by the mechanical structure (this accuracy is usually indicated by the manufacturer of the robot, and is typically ~ 0.1 mm). The angle range for link 1 is

$$\Phi = 2 \cos^{-1}\left(\frac{a^2 - b^2 + (x_i^2 + z_i^2)}{2ac}\right), \tag{7}$$

where a is the length of link 1, b is the joined length of links 2 and 3, and c is the distance from end-effector E_i to the origin (Fig. 31). The alternative positions of link 1 are

$$\text{positions of link 1} = \frac{\Phi}{\text{angular resolution}}.$$

The possible positions for link 2 (in our model) are limited. In fact, for all cases, apart from end-effector positions on the boundaries of the working envelope, there are exactly two available positions – an upper and a lower one (Fig. 32). The position of link 3 is uniquely determined by the first two links. Thus we conclude that for any given end-effector position within the working area, A_s is bounded by:

$$1 \le A_S \le \frac{4\pi}{\text{angular resolution}}.$$

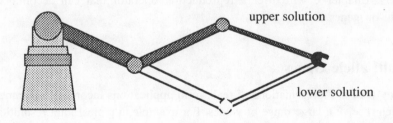

upper solution

lower solution

Fig. 32 - The two alternative positions of link 2.

The size of the trajectory space in our model can now be estimated by denoting A_s as the mean value for all end-effector positions, then the length of the path is 4.2 m, $n_{max}=20$, the boundaries around the desired path are 50 cm, the mesh is 1 cm^{-2}, and the angular resolution is 10^{-1}rad. These very conservative figures yield a trajectory size,

$$T_S = 2\pi 10 \sum_{n=2}^{20} \left(\frac{(50*(447 + 50)*1)}{n} \right) \approx 10^{90}.$$

It is clear from this example that the trajectory space for realistic trajectories, which incorporate hundreds of displacement vectors and a realistic positioning resolution, is monstrous in size.

Multi-bit, Multi-allele parameters

The identification of genes in chromosomes as well as their meaning in GAs is quite elusive. It is never quite clear where a gene starts and where it ends. Clearly, one has to abandon the notion 'one gene – one bit' for the trajectory representation. As we have already seen, the smallest functional unit is composed of several parameters, hence multi-bit genes [Greene and Smith, 1987; Shaefer, 1987; Smith, 1980]. Furthermore, the representation units, the arm-configurations, can receive a large number of values which suggests that the cardinality of the parameters is very high, hence multi-allele genes.

The idea of multi-bit genes as opposed to single-bit genes is an interesting representation format that can offer considerable flexibility for GAs. A minor disadvantage of such a gene structure is the need to modify the recombination operator so it shall only crossover outside these sub-structures [Suh and Gucht, 1987]. Chapter 6 will offer a reproduction operator that can accommodate multi-bit genes.

Multi-allele Genes

Many natural representations of real world applications incorporate parameters which receive a large range of values. For example, a typical joint resolution is one hundredth of a degree, enabling the joint to place its link in a few thousand different positions! We would prefer to use natural representations, but the schema theorem shows that a large range of alternatives in a single locus is counter-productive.

One way of solving the problem of multiple values is to have a sub-representation for each parameter, such as a binary sub-string whose length is determined by the range of the parameter values (see the ARGOT strategy in [Shaefer, 1987; Shaefer, 1988]). The sub-string approach is rooted in the school of fixed length representation ideology, but faces considerable technical problems such as overall string length and applicability of the conventional crossover operator. Nonetheless, it is a possible remedy, and for some problem domains it offers a satisfactory solution. We shall propose a different approach to multi-allele parameters which will reduce the number of allele values to a manageable size without compromising the solution's space resolution. We take the effort to introduce an alternative approach because it both solves this acute cardinality problem and offers an effective approach in many physical domains.

The idea is as follows: initially, the fine value range is replaced by a coarse one, spread at relatively large intervals, so each parameter has a relatively small number of alternative values. The solution space at this stage is therefore, of a low resolution. Allowing some time for the population to speciate to the degree the crude parameter values permit, the initial diversity of the parameter values is reduced to only a few selected values, if not fixed to a specific one. The algorithm assumes at this point, that each parameter value is fixed, and therefore, the reproduction cycle has degenerated. We shall call the assumed fixed value a *base value*.

When the search has thus been exhausted, the algorithm shifts its search resources to a new arena – the fractal arena. We develop a mechanism similar to some multi-grid methods. The parameter value resolution is increased, and the whole population is subject to a high rate of mutation. The sole purpose of the high rate of mutations is to increase the diversity of the population in the vicinity of the base parameter value. Because the diversity is concentrated around the base values, the search is now operating on limited sections of the solution space, the fractaled regions around the base values. The old base values remain unchanged since they are fixed throughout the population. The effect of reducing the parameter value interval is very similar to the effect of adding another decimal place to a real number, or the effect of adding additional digits (least significant bits) to a binary number. Because the new parameter values are of a smaller magnitude, there is no conflict between the base values and the new diversity [Jones, 1988].

This fractal resolution results in an algorithm that at any given time searches among few allele values, but ends up with a high allele resolution. The resolution of parameter values is increased to enable further specialization of the population members. This process of fixing the parameter values to base values, and continuing the search on a finer parameter value resolution can continue until the finite resolution, applicable to the system, is reached. The initial resolution should be sufficiently fine to allow the local maxima to be identified during the initial stages of the search, but it is a tricky business to foresee what resolution is adequate. As the search progresses, the resolution is increased at regions of high interest.

The cost of specialization achieved by fixing the parameters to base values and continuing with a finer value resolution, should be balanced against the robustness of the GA. In practice, it is not clear that the fractal resolution procedure locates, at intermediate resolution levels, the correct allele hyperplane that later leads to the optimum, or near optimum value. The additional premature convergence risk involved in this procedure is that at any intermediate resolution, the fitness of alleles may be misleading and 'deceive' the GA by favouring such values which do not represent the global optimum [Goldberg, 1987b; Goldberg, 1989a].

The fractal resolution encompasses a similar mechanism to that of Shaefer's ARGOT. Once a parameter value becomes fixed throughout the population, it becomes a base value located in the centre of a new value range. The new value range is of higher resolution and is typified by the decrease in magnitude of the previous value interval. Typically it is half the interval. The parameter can obtain any of the new values and, therefore, slide in either direction relative to the base value, but at a smaller value interval. Although the possibility of deception still exists on the allele level [Goldberg, 1989c], it is less of a problem in our specific domain due to the strong 'continuousness' of the joint space. Nevertheless, it should be noted again that the applicability of the fractal resolution is asserted here.

Searching for Hyperplanes in the Trajectory Space

Realizing the excessively large size of the trajectory space, it is interesting to find whether this space can be divided into sub-spaces of characteristic performance. In terms of GA processing powers, the question is: "Does the trajectory space contain hyperplanes?" The explanation of why GAs work and why they work so well in certain problem domains has been sketched in Chapter 2. The strength of GAs was attributed to the presence of hyperplane schemata – a space structure that can be recursively divided into structurally similar sub-spaces which have common characteristics. Unfortunately, the mathematical tools of schema theorem cannot always be extended to all representation formats. There are major problems in extending the schema theorem to flexible and position dependent representations. Nevertheless, the question whether the trajectory space contains hyperplanes, and what are its building blocks, is still interesting.

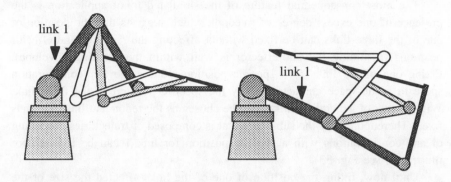

Fig. 33 - *Two trajectories, belonging to two different hyperplanes, which are aimed at following the same straight path. The position of link 1 is fixed for each trajectory, but differs in each case.*

The main importance of the schema theorem lies in its directive as to the type of effective building blocks [Goldberg, 1989b; Holland, 1975; Schaffer and Grefenstette, 1988]. The trajectory representation format is different from the ones the schema theorem was originally applied to, and an extension of the theory to cover such flexible representations is not forthcoming. However, the structure of the trajectory space can be analyzed (although less formally) to determine whether it exhibits features similar to those required by the theory. It is argued that an analysis of the representation space can be developed by investigating the building blocks that compose a schema, and the presence of hyperplane schemata in the trajectory space. If this space exhibits a schema structure and hyperplane schemata can be identified, then these properties can be translated into a representation format which will therefore, be suitable for a GA. This is a subtle argument that needs more support than is provided here. In the present section we shall advance an intuitive argument based on the correlation between the trajectory domain and the type of domains in which GAs are efficient. It is intuitive because it rests on engineering considerations rather than analytical analysis. By looking at the trajectory space, one can observe sub-spaces with a common characteristic, *i.e.* common both to the structure of the representation space and to performance. Such correlation between common representation features and common performance characteristics is an indication of the existence of hyperplane schemata.

The most consequential feature of the simulated robot application is the presence of one excess degree of freedom, which suggests that the position of one of the three links can be fixed without affecting the range of end-effector positions (assuming the end-effector is well within the working envelope). Fixing one of the links to a specific position reduces the arm-configuration space to a sub-space characterized by the specific position of the fixed link. Without loss of generality, link 1 can be chosen as the redundant link, and thus fixed. Thereupon, only the sub-space that is composed of trajectories consisting of arm-configurations with a specific position for link 1 can be formed (for illustration see Fig. 33).

Until now, fixing the position of one of the links affected the size of the trajectory space. Each fixed position of a link determines a unique sub-trajectory space similar to the way a schema defines a unique sub-string space. However, in order to establish a stronger analogy with the schema theorem, these sub-spaces of the representation space (genotype space) should have a correlating characteristic performance (phenotype space) so that the average performance of a unique set of parameters can be obtained (refer back to the example of counting schemata frequency in Chapter 2). For the robot application, the trajectory sub-spaces defined by a specific link position should exhibit a common performance characteristic. Indeed, such linkage between a specific link position and characteristic performance can be found in the simulated robot model. A specific link position results in a particular dynamic behaviour. More clearly, the position of a link influences the dynamic characteristics of the whole structure such as the global moment of inertia, the natural frequency of the structure, and so forth. Because the dynamic behaviour of the robot arm is what the GA aims to optimize, a specific link position can be associated with a characteristic dynamic behaviour. It is now possible to draw an analogy between the trajectory space structure and hyperplane schemata. The trajectory space can be divided into families of trajectories, which have both a common structure and a characteristic performance.

The sub-spaces described earlier can further be divided into more specialized sub-spaces, yet another desirable property of the trajectory space if the analogy between hyperplane schemata and a trajectory is pursued further. Link 2 can be constrained in various ways to form more specialized schemata. For example, by constraining the second link to either an upper, or a lower type, structure (as in Fig. 32). This type of constraint, as all other possible constraints on the two remaining links, have similar dynamic implications and constitute more specialized schemata.

So far, different trajectories have differed only in the arm-configurations that compose them (a fixed n value). However, a fundamental feature of the trajectory space is that n may vary. Can this additional representation aspect be linked to a characteristic performance and thus conform further with the schema analogy? The number of displacement vectors composing a trajectory has a clear effect on the global accuracy of the path-following program. In general (see Chapter 10 for a detailed explanation), more accurate trajectories can be created by incorporating more end-effector positions (more feedback for the control system during execution). Therefore, the trajectory space can be divided into sub-spaces characterized by the number of arm-configurations composing a trajectory, an additional schema character in the varying length trajectory space. The length of trajectories has only an indirect relation to the content of the trajectories and can be regarded as an independent characteristic of the trajectory space which is composed of sub-spaces of different trajectory lengths which in turn are further divided into sub-sub spaces of characteristic arm-configurations.

The hierarchical structure of characterizable sub-spaces, as just described for the simulated robot, is exactly what the schema theorem requires. Although the schema theorem cannot be applied literally to the trajectory representation format developed here, its essential structure can be traced in the representation. Further experimental investigation (Chapter 7) will provide additional evidence for the existence of schema structure and hyperplane schemata in this particular problem domain.

Over-specified Representations

Since GAs do not work directly with the problem domain, the representation of the problem is very important. In fact, GAs are 'unaware' of the problem domain because the algorithm only 'sees' the problem domain through the eyes of the representation which sits between the two and serves as the 'go-between' (Fig. 16). Therefore, information about the application which the algorithm cannot derive from the representation, will be excluded from the solution space.

The risk of excluding possible solutions by not representing the problem domain properly, is a more serious problem when constructing a representation for a real-world application due to the physical constraints imposed by the domain. In constraint applications, the parameters that best represent the problem domain are not always known to the programmer, while the natural control parameters of the process usually contain some degree of redundancy. To allow easier implementation of GAs to such domains, one should assume that the representation does in fact contain redundant information and equip the GA to handle such representations. Including redundant parameters in the representation leads to the following two questions:

(1) How does one incorporate such parameters in a representation?
(2) How does one resolve conflicts which may arise in redundant
      representations?

In Holland's discussion of broadcast languages and of the limitations a representation may impose on an adaptive plan [Holland, 1975], the possibility of incorporating additional parameters was introduced. Although broadcast languages were originally developed to create an evolving representation structure, they can be adapted to provide tools for the GAs to handle redundant parameters. It is important to note that for some applications, the excess information is artificially created, and exists only in the representation and in the algorithm's environment. The GA based on the idea of redundant representation structure has provided leverage to observe and analyze a fundamental process of adaptation in information systems, both natural and artificial. Chapter 10 is exclusively devoted to the aspects of redundant representations. Because machines frequently cannot work with under-specified or conflicting conditions, all conflicts resulting from redundant parameters must be resolved before issuing the control signals. In GA terms this means that the $p(g)$ function that translates the parameter string to the single decision variable d should filter the redundant information and provide an unequivocal phenotype (the use of phenotype is intended here to stress the mapping function that must be constructed when adopting a new representation format).

Including redundant information by way of additional synthetic parameters is very important to the robustness of a GA. In many cases, the machine's natural parameters are too few, or organized in such a way that it is very difficult for either the algorithm to identify co-adapted parameters, or for the operator to interpret the optimization results (not less important even if an offered solution works!). For example: the orientation of the tool held by the end-effector may require a specific configuration relative to the work piece, *e.g.* perpendicular. If the spatial configuration of a revolute arm is represented in the Cartesian format (three variables for the position of the end-effector and three for the orientation), then the GA can realize the linkage among the three orientation variables and cluster them together. If, however, the arm-configuration is represented in joint space, where the arm's position and orientation are determined by the complete set of links' position, then the unique spatial relationship between the end-effector and the work piece is hidden in all the joints and is therefore, much more difficult to extract.

It was suggested earlier that in many situations the programmer has neither the knowledge nor the time to construct an optimum representation. In order to relieve the programmer of the task of developing an optimum representation for a large class of applications, the representation should contain redundant parameters (provided the algorithm is equipped to handle the surplus information). Once furnished with the ability to work with over-specified representations, the algorithm can process redundant representations by identifying the most meaningful combination of parameters. This flexibility suggests a universal adaptability of the algorithm for each particular application. Identifying the more relevant parameters is called: *mapping the parameter space*, or *structuring the representation,* and is one of the important possible advantages in employing a GA. This was recently realized by a new model of GAs – a messy GA – proposed by Goldberg [Goldberg, *et al.*, 1989e]. Though redundant information makes the representation more robust, it should not be applied carelessly because additional information slows down the search. The use of redundant representations should be confined to situations when either the programmer is not able to suggest a better set of parameters, or when the increase in robustness justifies the risk of a decrease in efficiency.

An Example of a Redundant Trajectory Representation

Most robotic systems use the point-and-vector method to describe an arbitrary line in space. The point is denoted by the current end-effector position, and the vector is denoted by six variables defining a target end-effector position. Cartesian representation is one possible format to represent the vector for a target motion:

$$\mathcal{A} = \{X, Y, Z, \theta, \phi, \psi\},$$

where X, Y and Z represent the global spatial end-effector position relative to an arbitrary, but fixed, reference frame, and θ, ϕ and ψ (yaw, pitch, and roll, respectively) the orientation of the tool.

Whilst a Cartesian representation provides an immediate interpretation of the end-effector spatial position, it is disadvantageous in other respects. An arm-configuration of a non-Cartesian arm, an arm structure composed of some nonlinear links, can neither be uniquely determined nor directly controlled. This suggests that the kinematic parameters of the links are coupled in the representation space, which is an important disadvantage because GAs work better with orthogonal parameters. Representing a displacement vector in joint space might partially solve these problems. Thus, an alternative representation for the displacement vector is

$$\mathcal{A} = \{\alpha_1,...,\alpha_i,...,\alpha_d\}_{d=\text{dof}} \, ,$$

where α_i elements represent the position of the ith link. Realizing that the Cartesian and joint space representations have different advantages, it seems attractive to combine the two, and thus combine the best of both worlds. One possible way of incorporating the Cartesian information is to present it in two sub-sets, the spatial position (X, Y, Z), and spatial orientation (θ, ϕ, ψ) (the kind of information used by most robot programmers. Hence,

$$\mathcal{A} = \{\alpha_1,...,\alpha_d, (X, Y, Z), (\theta, \phi, \psi)\}_{d=\text{dof}} \, .$$

The Cartesian parameters are very useful in engineering applications. For example, high quality spot welding is achieved when a specific welder's relative orientation to the workpiece surface is maintained (usually perpendicular). While allowing a GA to choose between any Cartesian and link representation combination, the remaining parameters become redundant, and must be indicated as such to avoid confusion. One way of resolving the ambiguity of over-determined representations is to include markers or flags to indicate which of the multiple parameters have precedent [Goldberg, *et al.*, 1989; Goldberg and Smith, 1987; Holland, 1975]. Such flags can be thought of as hierarchical masking filters that enable/disable parameters to produce unequivocal spatial information.

Some additional support for this approach is provided in biology. In nature, there are DNA sequences that are not a part of a gene, but affect its expression [Ptashne, 1989]. These DNA sequences are called *activators* (or stoppers depending on their effect). In the arm-configuration representation structure given below, the string is composed of two distinct parts: the parameters and the promoters, $p(\ )$, which constitute the filter for the genotype expression (a similar representation structure was used to adapt crossover distributions [Schaffer, 1987; Syswerda, 1989]).

$$\mathcal{A} = \{((\alpha_1,...,\ \alpha_d,\ (X,\ Y,\ Z),\ (\theta,\ \phi,\ \psi)),$$
$$(p_1(\alpha_1),...,p_d(\alpha_d),\ p_{xyz}(X,\ Y,\ Z),\ p_{\theta\ \phi\psi}(\theta,\ \phi,\ \psi))\}_{d=\text{dof}}$$

In the representation model just presented, the number and nature of the redundant parameters is flexible, and is to be determined by the programmer according to the specificity required from the GA. The filter part of the string also exhibits some degree of flexibility, although it is greatly dependent on the parameter part of the representation.

Summary

This chapter opened with a statement about the robot model – that it should not represent any specific robot system. Rather, the desired features of the model should focus on the inherent problems of a robotic environment in so far as they are common to general robot trajectory generation procedures. Several definitions concerning trajectory generation and trajectory representation were introduced to simplify the discussion about the robot model. With the definitions we also analyzed the complexity and size of the trajectory space. It was discovered that even for very conservative path lengths, the size and complexity of the trajectory space is considerable. Complexity in our context means that conventional optimization techniques may be precluded for most realistic applications.

Up to this stage, we represented trajectories in their natural format because this format is universally applicable to all robot systems. Unfortunately, this format is unsuitable for a conventional GA. After analyzing the trajectory space it was suggested that it might be more constructive to modify a GA to suit the trajectory structure than it would be to develop a mapping function to convert the natural structure into the rigid structure of conventional representations. Consequently, after deciding to venture into the natural representation, an analogy between the representation of the robot trajectories and schema theorem, was discussed. It was argued that it is not necessary to adopt the standard GA representations and operators in order to construct a successful GA for trajectory generation.

The last section of this chapter was speculative in nature. It proposed an over-specified representation as a suitable representation to employ when applying a GA to a real world process. Redundant representations are viewed as a major characteristic of natural systems and a prerequisite feature of general adaptive systems. Chapter 10 will be devoted exclusively to this topic.

Chapter 6

THE ALGORITHM FOR
TRAJECTORY GENERATION

Out of the large mass of problems that exist in applying GAs to a class of trajectory generation applications, we introduce a GA which successfully handles the trajectory generation of the redundant robot model described in Chapter 5. The various GA operators used in this model will also be defined. In this chapter the actual algorithm is presented together with the algorithmic solutions to the various problems we encountered in applying GAs to trajectory generation. As could be predicted, the main modifications to the standard GA model take place in the recombination part of the algorithm. The first section of this chapter is devoted to that mechanism.

A general comment regarding this chapter: although the focus of this book has been on trajectory generation, it was suggested that trajectory generation is really an instance of a broader family of applications – the domain of order dependent processes. We do not intend to branch out at this stage to other models, and shall focus on an algorithm to solve the trajectory generation problem. However, it is believed that the reader will gain insight into the working of the GA if the more general issue of an order dependent process is considered.

Trajectory Reproduction

Before the workings of the modified GA are investigated, an important change in terminology should be defined. *Reproduction* in this book refers to the whole mechanism by which new genotypes are created. This specific use of the term is in conflict with the common use of the term reproduction in GAs literature which has come to designate the first stage of recombination – the duplication of a string before crossing it over with another string.

In traditional recombination operators, after choosing a cross site in one parent genotype, the corresponding cross site in the second parent is readily determined according to the locus of the cross site in the first parent. Hence, corresponding cross sites are situated in the same positions in their respective genotypes. The criterion for matching cross sites is the position in the genotype. This type of cross sites matching criterion can be associated with the biological term *homology*. The use of the word *homology* here refers to the fact that conventional string representations have corresponding structures when crossed. This contrasts with the biological use in which *homologous* organs are body organs of different species sharing the same evolutionary origin that have come to serve different purposes. (The members of each pair of chromosomes are called *homologous chromosomes*, and exhibit a specific length, centromere placement, and identical gene sites (or loci) along their axes.) Similarly, *analogous organs* are organs or parts of the body of different species that have different evolutionary origins, but have become adapted to serve the same purpose.

Homologous cross sites are not suitable for our trajectory representation due to the weak correlation between the sequential position of an arm-configuration, and the corresponding end-effector spatial position (its function in the trajectory). The varying length and order dependency of trajectories destroys the role of position with which we are familiar from traditional GAs representation. The order dependency also prevents the use of scaling operators to normalize the string length (such as the scaling effect of the dynamic time warping used in speech recognition [Vintsyuk, 1971]).

Analogous Crossover

We introduce a new matching criterion for the trajectory model – an *analogous* criterion for matching cross sites – a cross site correspondence criterion based on similarity of the genotypic character of the arm-configuration at the cross site locus. In the general case of order dependent processes this will amount to matching genotype elements according to their similarity in some parametric metric space. Matching parameters for crossover according to their genotypic character rather than according to their genotypic position is a fundamental modification of the conventional recombination operator. After choosing a cross site in one parent string, the corresponding cross site in the second parent, is determined according to the proximity of the arm-configuration at the locus to the arm-configuration space – the predicted phenotypical feature (henceforth, similarity in the phenotype function emphasizes the functional characteristic of the analogous crossover). The locus of the cross site of the first parent has neither significance, nor a role in determining the locus of the corresponding cross site in the second parent.

Matching parameters according to their phenotype function can be illustrated by a road system. Driving between two arbitrary points, A and B, one can devise several alternative routes, some of which may even have shared paths. The economical way of combining different routes is to switch over either at shared stretches of road, or at intersections of routes. However, at some instances the cavalier driver might venture switching to another route that does not meet the one he is currently on. If the aim in trying different routes is to minimize travel distance, then switching routes that do not overlap is more likely to occur at positions where the routes are in close proximity to each other, thus minimizing the time wasted on the switch.

The analogous cross site criterion not only preserves the 'orderliness' characteristic of the strings, but also has a biological motivation and intuitive justification (such as in the above road travel illustration). In a complex string structure where the number, size and position of the parameters has no rigid structure, it is important that the crossover occur between sites that control the same, or at least the most similar, function in the phenotypic space. The extent of 'unfixedness' of the modified representation which caused the precise locus to have a diminished significance, makes a mechanism like the analogous cross site a necessity. Of course, there are many phenotypic similarity criteria one could use in an analogous crossover mechanism. Here the end-effector position was chosen, but another possibility might be the similarity in the structure of the whole arm-configuration (a metric distance in arm-configuration space). We chose a simple matching criterion for the trajectory model – the Euclidean distance between end-effector positions – because it is simple to compute and is unequivocal. Defining a similarity measure between the whole arm is more complicated than one might at first suppose (*e.g.* the position of the first link is significantly more important than that of subsequent links). In any event it is liable to be computationally expensive. Whatever criterion one adopts, the analogous based crossover provides the least disruptive crossover operation.

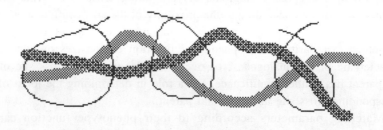

Fig. 34 - *Two paths resulting from two different trajectories. The circled sites are preferable for crossover in order to minimize path disruption.*

It is obvious from Fig. 34 that it is least disruptive to crossover at the regions where the two paths meet, a site which constitutes an attractive site for crossover. However, to crossover at phenotypically similar sites requires a mapping function from phenotype to the genotype, an alien requirement to fulfill for most biological systems, and computationally demanding for many artificial systems. In trajectory generation and sequential processes in general, the correlation between the phenotype and the genotype is simpler than in parallel processes (such as protein folding), but still only in rough approximation.

The difference between the two cross site matching criteria described (as would be for any criterion for analogous crossover) runs deeper than just the difference in computation complexity. Because the criterion is based on functional aspects of the representation parameters, the criterion has a physical meaning which should be analyzed. For example: matching overall similar arm-configurations means that the changeover at the cross site shall be the least disruptive from the arm motion viewpoint (minimum jerk in joint space), but does not address the resulting end-effector position directly. Therefore, we may have two arm-configurations that are most similar to each other, but are not the most similar in their end-effector position. Because ultimately we are concerned with the path being drawn by the end-effector, the advantages are clear in specifying crossover directly in terms of the end-effector.

Segregation Crossover

A subtle feature of the crossover operation that was not visible when only single-bit parameters were used becomes relevant in the presence of multi-bit parameters. In nature, crossover can occur either between genes, or within genes. In order to differentiate between the two alternative cross sites, the crossover operation which crosses between genes is called *segregation crossover*. In conventional GAs representation, where a parameter is composed of only one bit, cross sites cannot split a parameter, and the two crossover mechanisms degenerate into one mechanism – crossing over between parameters. When multi-bit parameters have a natural meaning, such as in process control applications, crossing over within those groups may have a disruptive effect on the genotype structure and consequently the phenotype function [Greene and Smith, 1987; Shaefer, 1987; Smith, 1980].

Segregation recombination (crossing between genes) is the only crossover mechanism used in this work, and proceeds according to the following sequence (Fig. 35 gives the flow diagram):

(1) Choose two parent trajectories <u>probabilistically</u> according to fitness.
(2) <u>Randomly</u> choose a cross site in one of the parent trajectories.
(3) Slide the second trajectory along the first in order to find the most analogous end-effector position to the end-effector position at the cross site of the first parent.
(4) Repeat steps 2 and 3 for a second cross site from the first cross site onwards.
(5) Copy the counterpart sub-trajectories defined by the two cross sites from the two parents to obtain a new offspring trajectory (only one offspring trajectory!).

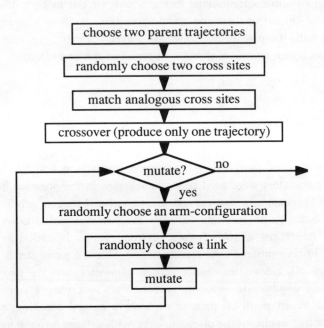

Fig. 35 *- A flow diagram of the recombination routine.*

Fig. 36 shows a compressed version of two trajectory genotypes (compressed in the sense that only the end-effector positions are displayed for each of the arm-configurations composing the two trajectories). The two randomly chosen cross sites are marked in one of the parents, while the analogous matching points are shown in the second parent. The crossover is a 'crossing after' operator. The resulting offspring trajectory (Fig. 37), whilst containing information from both parents, is different from both of its parent trajectories in the number of arm-configurations comprising it.

It has to be stressed again that the end-effector's space is an equivocal trajectory representation in general, but especially when redundant degrees of freedom are incorporated in the structure. The mapping, from end-effector space to arm-configuration space, is highly nonlinear and discontinuous (Fig. 26). By matching cross sites with the end-effector proximity criterion one inherently introduces an ambiguity into the recombination operator, an ambiguity which results from the difference between the end-effector and arm-configuration spaces.

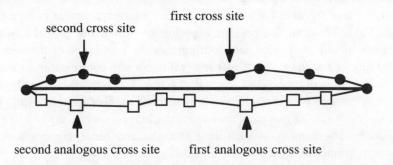

Fig. 36 - *The desired straight line path (movement from right to left) and two end-effector trajectories. The two cross sites of the first string, and the analogous matched cross sites in the second string, are indicated.*

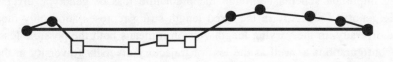

Fig. 37 - *The trajectory resulting from the crossover shown in Fig. 36.*

The Mutation Operator

The mutation operator is largely a background noise, a subtle protest against the focussing effect of the recombination operator. GAs for unfixed length strings have to control the premature loss of alleles and the string length diversities. The use of analogous matching, and the dynamic effect it has on string lengths and arm-configuration sequences, helps to reduce the danger of premature allele loss in the way arm-configurations are interchangeable. An allele at one locus in one trajectory can be at a totally different locus in a different trajectory.

The mutation operator should not be 'concerned' whether the string is composed of single-bit or multi-bit parameters. An occasional change can occur at any locus in the string. For binary parameters, the extent of the mutation is readily determined (flipping the bit value between the two possible values). Where multi-bit parameters are present it is arguable that mutation should be less probable for bits which control a large parameter change (high order bits). Thus, in the present experiments whilst the probability of a mutation at any particular arm-configuration is uniformly distributed, the probability of mutating a particular link within an arm-configuration is inverse to the link's significance (link 1 : link 2 : link 3 :: 1 : 2 : 4). Finally, each mutation changes the parameter value by a round number of units, while each unit has the value of one parameter interval (decreases when the resolution is increased). The number of such units changes, and hence the extent of the change is controlled by a Poisson distribution having $\lambda = 0.3$.

Deletion and Addition Operators

In simple GAs, the recombination operator intrinsically involves the risk of losing important schemata through the premature loss of genotype diversity. While genotype diversity in the fixed length string representation only means allele diversity, in the varying length version it means both the possible loss of arm-configurations as well as the loss of trajectory diversity (diversity in the n space). Some algorithmic measures are necessary to control the dynamics of the varying string lengths. The analogously driven recombination partially works towards this end, but the algorithm still requires operators that work directly on the string length. These operators are called *addition* and *deletion*.

In order to improve the stability of the strings' length, two additional operators, deletion and addition, are included in the reproduction mechanism, and provide a background noise over the string lengths. The deletion and addition operators are new operators to the simple GAs, but neither new to nature, nor to the GA theory. The mechanics of the deletion and addition operators are very simple, and very similar to the mechanics of the mutation operator. A random change in a string length is imposed by either deleting an arm-configuration of the string, hence deletion, or by adding a new arm-configuration to the string, hence addition. The deletion operator simply chooses at random an arm-configuration to delete (see Fig. 38 for illustration). For the addition operator, several policies to determine how to add an arm-configuration are possible. It is not so critical which policy is used as it is more dependent on the specific application. We consider here three schemes for the addition operation: duplication, related, and random.

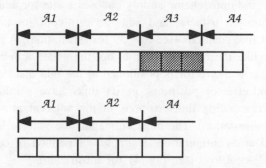

Fig. 38 - *An illustrated deletion operation of a complete arm-configuration ($A3$ in this case).*

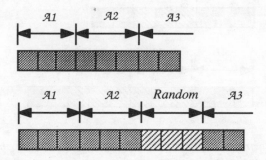

Fig. 39 - *Illustrative random addition of an arm-configuration.*

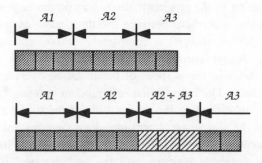

Fig. 40 - *An illustration of a related addition. The structure of the added arm-configuration is based on its adjacent arms.*

The random addition scheme simply chooses a site for addition (between two arm-configuration triplets) and adds an arm-configuration *de novo,* a random triplet of link positions (see Fig. 39 for illustration). A related addition scheme operates like the random scheme, but the added arm-configuration is such that either its end-effector is positioned at the mid distance between the two adjacent end-effector positions, or its links have a mid metric value between the corresponding link positions in the adjacent arm-configurations (see Fig. 40 for illustration). The most unimaginative scheme is the duplicant where the added arm-configuration is simply a repetition of one of the arm-configurations adjacent to it (see Fig. 41 for illustration).

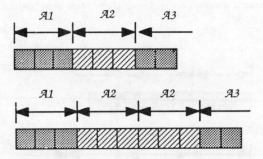

Fig. 41 *- An illustrated duplication addition operation where the second arm-configuration was duplicated.*

In similar fashion to the pragmatic discussion on the analogous crossover, we shall analyze the differences between the three schemes for addition. The random scheme for addition is a completely 'naive' scheme introducing completely new information to the arm-configurations pool. This is advantageous in terms of the purpose of the mutation operator, but it is 'too advantageous by half'. The introduction of a random arm-configuration carries a high probability of disrupting the stream flow of the trajectory especially at advance stages of the search, *i.e.* it will introduce vigorous link displacements between its adjacent arm-configurations. The related scheme has a smoothing effect (usually a desirable property) on the path because it increases the sampling resolution by interpolating between end-effector positions. Both from control and engineering considerations the related addition makes a lot of sense and potentially offers an important optimizing operator. However, it is difficult to justify it biologically. There is very little evidence for such directed and 'intelligent' translocations of DNA sequences. If it is just a one-off or purpose dedicated optimization one is after, then the related addition is a plausible scheme. However, for the general investigation of adaptive systems based on natural mechanisms we have to be prudent when incorporating non-biological operators. Pragmatically and for reasons as set above, we shall restrict our algorithm to use only the addition operator of the third type – the duplicant addition operator.

The duplicant scheme is a rather intriguing one. On first impression it appears to offer very little, it does not change the phenotype and therefore does not change the performance of the trajectory, and it does not introduce new 'genetic' material to the population (new arm-configurations not present before). Besides increasing the sheer quantity of arm-configurations which its bearer trajectory contains, it does not change anything. However, this state of affairs is only temporary. Once the twin arm-configuration is separated due to a crossover, the 'dormant' arm-configuration becomes active. This is a simple mechanism through which individual arm-configurations can proliferate and propagate through the entire population.

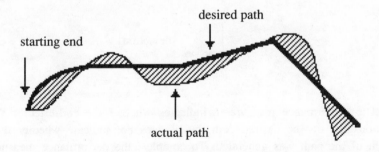

Fig. 42 - *An illustration of a desired path (heavy line), the actual path drawn by the end-effector when following a given trajectory (winding line), and the accumulated error from the desired path (diagonal columns).*

The Evaluation Mechanism

The evaluation of trajectories, as all evaluation functions in GA applications, is external to the GA. The trajectory created by the reproduction mechanism is sent to the simulation package which, while executing its motion commands, returns a time based vector of triplets of link positions that were visited while the robot arm was in motion (20 Hz sampling rate). This vector is a reflection of the dynamics of the arm and the position of the end-effector during motion (see Fig. 28 for illustration). The actual path (the phenotype in our model) can be assessed in a variety of ways determined according to the goal function, hence the goal of the optimization. In our model the goal was to minimize the total deviation from the desired path, but other practical goals may be minimizing the maximum deviation from the desired path, minimizing the ripple of the path, and so forth, including any combination of such features.

The output from the evaluation operator is a positive number indicating the relative fitness of a phenotype to the other members of the population. The fitness is based on the absolute performance of the trajectory. In the trajectory path-planning, performance d_t is the accumulative (total) deviation between the actual and the desired path (Fig. 42), and is obtained by

$$d_t = \int_{\text{trajectory}} |\text{deviation}| . \tag{8}$$

The performance measure d_t indicates whether the end-effector visited positions not on the desired path, but it does not indicate whether the full length of the path was generated. To complete the performance measure we add another performance feature l_t which measures the discrepancy between the desired starting and terminating ends of the desired path and the corresponding ends in the actual path. Hence,

$$l_t = |(E_{\text{desired}})|_{\text{begin}} + |(E_i - E_{\text{desired}})|_{\text{end}} . \tag{9}$$

The complete performance measure therefore becomes,

$$d_t = l_t + \int_{\text{trajectory}} |\text{deviation}| . \tag{10}$$

The fitness of a trajectory is a relative measure which depends on the performance of other members of the population. The fitness is a positive monotone transformation of d_t designed to give an appropriate relative reward to different members of the population. An exponential scaling is used so that

$$\text{fitness} = 100 * \exp^{-d_t} . \tag{11}$$

In order to maximize the resolution of the fitness space, a scaled and normed performance measure d_n replaces d_t in the following way:

$$d_n = \frac{d_{\max} - d_t}{d_{\max} - d_{\min}} . \tag{12}$$

Although the total deviation is a measure used to calculate the fitness of a trajectory (a single decision variable), the detailed information how deviation was accumulated is not lost and is recorded in the trajectory in the way of a deviation profile. This deviation profile has no role in normal GA applications, but will prove essential when sub-goal reward policies are implemented (Chapter 8).

Two general observations are worth mentioning. First, if fitness is heavily influenced by some small number of parameters and lightly by the rest, then the efficiency of the GA search is hampered. In an extreme case, where fitness is predominantly influenced by one parameter, the algorithm is wasting search resources on parameters which do not contribute to a worthwhile fitness (similar to a search in a flat fitness landscape). This is a theoretical issue of selection and adaptation which becomes important in the trajectory application, and is likely to be important in many other real world applications. Chapter 8 is devoted to this subject exclusively. The second important observation is that our fitness function used exponential scaling. Generally, nonlinear scaling potentially contains a risk of introducing bias into a GA [Crow and Kimura, 1970; Westerdale, 1989]. The argument goes as follows: nonlinear transformation from the performance (which is absolute) to the fitness space (which is relative) clearly changes the topology of the surface the GA is climbing on, but may change it in such a way that a misleading bias is introduced against the original peaks. Another, and more intuitive, argument is that too much scaling may boost the probability of premature convergence.

The justification for using such a powerful scaling function here is founded on the following argument. We are interested in a practical method with which a good trajectory can be located among an excessively large number of plausible trajectories. It is not good enough for our purposes that the GA may be more efficient than all other known optimizing algorithms if the number of evaluations it requires hinders practical implementations. We want meaningful improvements in performance and we want them fast! Scaling is an important subject which is still unresolved in the literature and deserves much more attention.

The Selection Mechanism

In simple GAs, good control over premature loss of diversity is achieved by adopting a replacement scheme that inherently establishes a controlled convergence and a proportional distribution of search resources according to relative fitness, and therefore works against the exclusiveness of a particularly fit solution arising stochastically at intermediate stages of the search [Booker, 1982; Cavicchio, 1970; De Jong, 1975; Goldberg and Richardson, 1987]. The importance of achieving a controlled evolution is that the merits of the genetic operators can be fully exploited only when they are given free rein and are not used to balance the effect of each other (such as inept attempts to avoid premature convergence by boosting the mutations rate).

The implementation of sharing or crowding functions, either phenotypic or genotypic, is unfortunately impractical. It is difficult to concoct a similarity measure for isomorphism in the genotype space due to the presence of varying n and the redundancy incorporated in the structure, and the order dependency of arm-configurations. Furthermore, it is not clear how a phenotypic isomorphism is computed, and most methods are likely to be computationally prohibitive. As a result of these difficulties, but still desiring to employ some restrictive mechanism on the proliferation of stochastically highly fit trajectories, we adopt the *preselection* mechanism [Cavicchio, 1970]. The preselection mechanism is a scheme in which an offspring replaces its parent.

We shall replace a parent only when its offspring has a higher fitness. Because offspring are just a recombination of the information contained in their parents (something like the musical meaning of a 'variation on the theme'). Broadly speaking we do not lose information when we choose any one of the generations (the parent or the offspring). However, when we choose the one with higher fitness we choose a generation in which the information is better arranged. We deviate slightly from the above paradigm with the trajectory model because we have only one trajectory and two parents. So we choose the best two among the three to remain in the population.

One aspect of selection we introduced in this model was the artificially imposed parameter fixation to base values and the subsequent high mutation rate. We shall now analyze how it is implemented by going back to our natural representation. At any given stage, the number of different link positions for any link is set at 10. Initially, each link has 10 joint positions equally spaced at $2\pi/10$ rad intervals. When the algorithm reaches a saturated population, and most of the arm-configurations are composed of a small number of link positions, the best trajectory is kept, and the rest are discarded. The reduced population is duplicated to reach the original population size while superimposed with a multitude of mutations of half the magnitude of the previous interval size. The process of fixing the base value by choosing a single trajectory and replicating it with high probability for mutation, is a kind of hybrid process between a GA and a multi-grid method. In the following chapter we shall see the effect this mechanism has on convergence and specialization.

Summary

In this chapter the GA for trajectory generation – trajectory-GA – was defined. The proposed GA incorporates few new mechanisms which are necessary to make a GA amenable to the natural trajectory representation we chose to explore. The majority of changes which were needed concentrated in adopting the reproduction operator to suit the varying in length, and an order dependent representation of trajectories. We defined an analogous crossover operator which uses a functional matching criterion to match cross sites between the parent trajectories. Two new operators, addition and deletion, were introduced. These new operators provide a background noise to the number of arm-configurations used to define trajectories. A suitable mutation operator for the trajectory representation was also defined.

The evaluation mechanism used to evaluate the performance of trajectories was defined. The fitness function used here is relative, meaning that the fitness of trajectories is scaled. The reasons for using an exponential scaling function were posed against the drawbacks, namely to speed up convergence at the price of risking premature convergence. Finally, the selection mechanism was analyzed and an explanation was offered as to why it is preferred in this work to use a preselection mechanism where offspring enter the population only by replacing their parents, and only if they have higher fitness.

Chapter 7
EXPERIMENTAL RESULTS

In Part I of this book we discussed the adaptive mechanisms of evolution, and introduced an algorithmic framework with which these mechanisms can be reproduced in an artificial environment. In Chapter 4 we analyzed the robotic environment, an environment (and particularly the generation of robot trajectories) that was recognized as one which may benefit greatly from an efficient optimizing algorithm. We also discovered that the intrinsic characteristics of trajectory generation often become too complex for conventional optimization algorithms. Consequently, we suggested that a GA might be a rewarding approach to the optimization of trajectory generation. We further suggested that due to the rigid structure of standard GAs, there is a wide interest to develop a GA which will process a natural representation structure rather than develop some auxiliary representation for trajectories. We then discussed the plausibility of developing such a GA. In Chapter 6 we defined a GA for trajectory generation which we shall name *trajectory-GA*. Having laid the groundwork, we are now ready to apply the 'proof of the pudding' test to the trajectory-GA and see whether the many modifications we introduced to the standard GAs practice do indeed work. This we shall do in the present chapter.

This chapter shall, therefore, be devoted to experimental tests with the trajectory-GA. The path specifications we chose are the following: path length of 4.2 m, with its beginning situated at the position where the robot arm is fully stretched to the right, and its other end at the location (-1, 1) [m], relative to the centre of symmetry of the first joint (Fig. 4). The three links are free to rotate around their respective joints, but are restricted to the vertical plane. There are no mechanical limits restricting the positions any link may adopt. We have chosen the following parameters for the GA: the population size is set to be 100 trajectories, which are initially created at random choosing $n = 10$. At any given time, each link can be positioned at 10 different positions. The initial population is created by randomly assigning a link position to each link of an arm-configuration. This process is repeated for all arm-configurations except for the first one of each trajectory which is always the fully stretched position to the right (the first arm-configuration was fixed for purposes of convenience and is arbitrary). Mutations, deletions, and additions occur only after a recombination ($p_c = 1.0$) and typically have 0.05 probability. The mutant link position is superimposed with an angle change equal to whole link intervals whose number is determined by a Poisson distribution with $\lambda = 0.3$. The addition operator has a slightly greater probability (≈ 0.06) to balance the advantage that trajectories composed of a small number of arm-configurations have at the initial stages of the search (this point will be discussed in detail in Chapter 10).

The increase in link position resolution is determined according to the population diversity in the phenotype space (the difference between the average error and the best error). On average an increase in resolution occurs every 400-500 steps. The increase is irreversible (in contrast to the ARGOT strategy [Shaefer, 1987]). When the GA 'decides' to increase resolution, only the best trajectory is preserved and it is duplicated to replace the entire population. A high rate of mutations ($P_m = 1.0$) is applied during the duplication phase (one trajectory is protected from the mutations to ensure continuity in the best result curve).

Performance of the Trajectory Genetic Algorithm

We first set out to examine whether the trajectory-GA works at all. Fig. 43 plots the best results from five different experiments, the trajectory-GA being

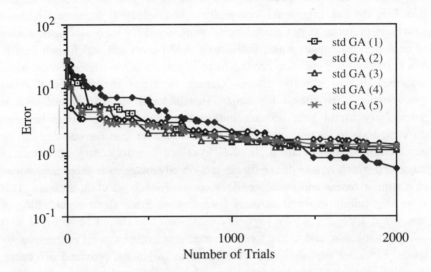

Fig. 43 - Best string for three trajectory-GA experiments.

assigned to optimize the same path, but each time following a different random numbers sequence. We can observe that the algorithm indeed optimizes trajectories and that the rate in which improvements are introduced is impressive (we shall later compare the performance of the trajectory-GA with other algorithms). Furthermore, we can also observe that in general the algorithm is quite robust and repeatable (the standard deviation is about 20%). It should be noted that the ordinate has a logarithmic scale.

Although the best trajectories at the end of each experiment are different, their performances are of a similar quality. This phenomenon indicates that the trajectory-GA settles on different regions in the trajectory space. This seemingly unstable behaviour of the trajectory-GA can be expected and explained when the error of the best trajectories is re-examined. The errors show little diversity. The phenomenon of settling on different peaks in the genotype space, while fitness diversity is relatively small, is called *genetic drift*. Genetic drift is recognized to be an inherent feature of GAs which do not incorporate a sharing function (or similar mechanism) [Goldberg and Richardson, 1987]. In general, genetic drift is known in biology, and is a frequent phenomenon in population genetics where different individuals within a species or different species do not compete for the same resources and therefore do not experience competition and selection pressure. In these situations evolution is not guided by adaptation and the mere stochastic nature of evolution determines which individuals shall persist and which shall vanish. We know that the selection driving force is not the actual characteristic of an organism, but its effective fitness. Genetic drift can also be viewed as an evolution in a flat fitness landscape. This pictorial view of genetic drift is particularly useful here because indeed GAs cannot differentiate between different trajectories once these have an error smaller than the one that can be detected by the simulation. In the 'eyes' of our trajectory optimizer, a trajectory which has a similar fitness has no advantage over other trajectories of a similar fitness and therefore, does not experience selection pressure. This is an opportunity to recall an early comment we made about optimization of real world applications. We suggested that often one should be satisfied with one good solution, and not be too concerned about other aspects of the solution space, including the possibility of even better solutions, provided of course, that the solution at hand is good enough. (In GA applications where it is important to maintain genotypic diversity, some special niche and species operators can be considered [Cavicchio, 1970; Deb and Goldberg, 1989; De Jong, 1975; Goldberg and Richardson, 1987].)

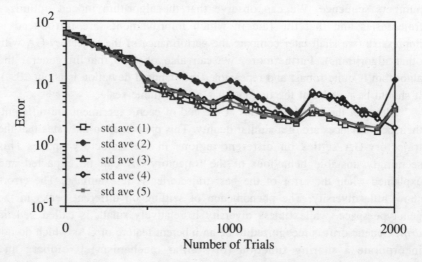

Fig. 44 - *Average error as a function of search progress of three trajectory-GA experiments.*

We shall conduct a similar analysis on the average fitness of the whole population during the five experiments. Fig. 44 plots the average fitnesses of the five experiments presented in Fig. 44. In general, one would look for a similar behaviour among different tests and regard such similarity as an indication of robustness. We observe two things. First, the curves are neither smooth nor monotonous (as one might have expected in studying most GA applications reported in the literature). Second, the average fitness curves exhibit a similar behaviour even though they follow different paths (initially they almost overlap). The drastic 'jumps' in the average fitness value are a result of the special selection mechanism we introduced and these jumps will be explained in the following paragraph. We should note that if one does accept the jumps, the behaviour of the curves has the characteristic convergence behaviour.

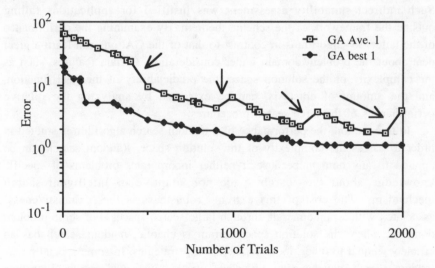

Fig. 45 - *Best and average errors of one trajectory-GA experiment. The four arrows point at a resolution change.*

Finally we turn to investigate in greater detail the evolution of one trajectory optimization. Fig. 45 plots the best and average fitnesses of a population in a typical experiment. The four drastic changes in the average error (indicated by the four arrows in Fig. 45) indicate the point at which an increase in joint angle resolution occurred. The change in the average error at these sites is the result of two contradictory processes: one which reduces the average fitness, and one which increases the average fitness. The decrease is a result of the reduced trajectory diversity due to focusing on the best trajectory. The increase is due to the high rate of mutation induced at this stage. Focusing on one trajectory has great effect at the initial stages of the search, and a lesser effect as the search advances. Therefore, the average error will tend to decrease at the first resolution change, but thereafter increase during subsequent changes.

Between Random and Genetic Search

In the discussion about the suitability of a given representation to a GA processing (Chapter 2 and in greater detail Chapter 9), the performance of a GA was compared to those of random search and hill-climbing algorithms. Such indirect suitability assessment was justified for applications falling outside the framework of the schema theorem. By examining the performance of the latter two algorithms in contrast to that of the GA, one can learn a great deal about the problem domain under consideration. Certain features, such as the complexity of the solution space, the predictability of the goal function, and the amount of epistasis, can be extracted by analyzing the relative performances of the three search procedures.

In this section, the performance of a random search algorithm is studied in order to assess the complexity of the solution space. Random search can be applied to any domain because it neither incorporates problems of specific knowledge about the search space nor adaptive or intelligent search mechanisms. The cost of this extreme robustness is the 'exhaustiveness' associated with a random walk through large search spaces. For some problem domains, where the solution space terrain is chaotic, random search has an efficiency equal to other, more sophisticated, strategies. In some special cases, random search may have an advantage over sophisticated search algorithms since it cannot be misled to focus on a sub-optimal region of the solution space and thus irreversibly converge on a sub-optimal solution.

The expected performance of a random search can be calculated in the following two ways:

(1) By estimating the error density function associated with the trajectory space.
(2) By stochastic trials.

The expected error density is obtained by selecting trajectories at random, and investigating their error distribution (Fig. 46). Assuming the error distribution is normal by nature, the estimated error density curve may then be computed (Fig. 46). The hypothesis that the trajectory error distribution is normal is verified by a x^2-test with the following parameters: $x_{Ave}= 65$, $\sigma = 17.5$ and $\alpha = 5\%$. The hypothesis that the error distribution is normal is accepted on the basis of the good fit of the initial population. By establishing the characteristics of the error density curve, we can obtain the probability of locating a trajectory at random which has an error less than a given value. This probability indicates the relative performance of the trajectory-GA.

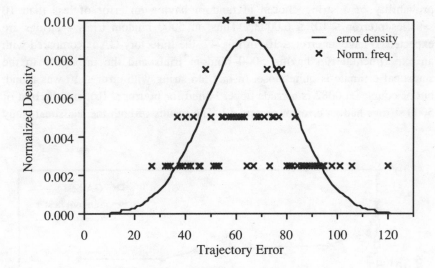

Fig. 46 - *The error distribution of a randomly selected population comprising 100 trajectories.*

A subtle point of a normal distribution is that errors smaller then zero have a non-zero probability. When comparing the relative performance of a random search with that of a GA, one is interested in the error region close to zero. The nature of a normal distribution, and its approximation, makes predictions at the trailing edges of the curve (in this case the regions outside $\pm 3\sigma$) relatively unreliable. To this effect, the nature of a goal function adds an additional distortion – the fact that fitnesses and performances are always positive, bounded by zero. For example: according to the error density curve the probability of locating a trajectory with an error less than one is p(error ≤ 1) = 0.0002 while the probability of locating a trajectory with an error less than or equal to zero is p(error ≤ 0) = 0.0001. While the former error is applicable, the latter is not. The proximity between the two probabilities indicates that the error density curve is not accurate in the region close to zero.

A stochastic experiment provides a quantitative verification of the performance estimated by the error density curve). The probability of locating a trajectory with an error less than 15 is p(error ≤ 15) = 0.0023. This agrees with the experimental results in Fig. 47. As described above, p(error ≤ 1) is between 0.0002 and 0.000. From this analysis one can conclude that the probability of a string chosen at random having an error of less than 10 satisfies p(error ≤ 10) ≤ 0.00082. Thus, in 5000 random trials 4 strings are expected to have an error ≤ 10. In Fig. 47, the trajectory-GA is compared with an experimental run having 5000 random trials and the proximity to the statistical estimate is quite close. In fact, no string with error ≤ 10 was found, but of course 0.00082 is a crude upper bound for p(error ≤ 10), and the best of 5000 strings had an error of around 12. The results of both the statistical study

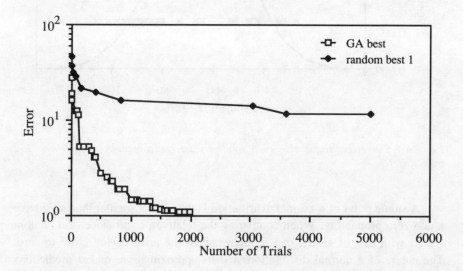

Fig. 47 *- Relative performance between random search and GA.*

and the stochastic experiments suggest that the trajectory-GA has a search efficiency which is better by two orders of magnitude than that of a random search.

Hill-climbing Versus Genetic Algorithms

The performance of a hill-climbing algorithm is compared with that of the trajectory-GA. The performance of a hill-climbing algorithm indicates how complex the solution space is. The results of four hill-climbing experiments, starting from a randomly chosen trajectory, are presented together with a typical GA performance (Fig. 48).

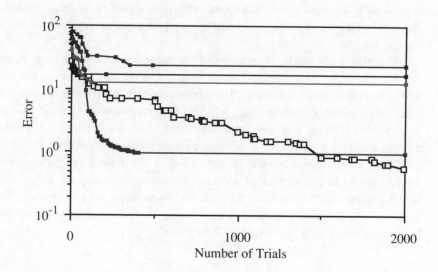

Fig. 48 - *Results of four hill-climbing experiments (solid squares), and of a typical GA (hollow squares).*

The large diversity in performance of the hill-climbing algorithm indicates that the trajectory space is multi-modal, and that it is unstable, *i.e.* prone to settle on a sub-optimal solution. The multi-modal character of the trajectory space is further substantiated when the genotype trajectories of the final results are analyzed. The final trajectories have few parameter values in common.

This fact can also be observed in all final trajectories arrived at by the other search trials. Compared with the performance of the hill-climbing algorithm, the performance of the GA is much more predictable and consistent. The fact that the GA also settles on different genotype regions does not diminish its reliability since the GA is guided by the fitness, and as such there is little difference among the different experiments.

Summary

The most important question this chapter set out to answer was the question of whether the trajectory-GA works. The results presented demonstrate the power and robustness of the trajectory-GA. The new operator introduced by the order dependent GA model – the analogous crossover – proved to be a crossover operator which is capable of handling order dependent representations of varying lengths.

The efficiency of the trajectory-GA, compared to that of random search and hill-climbing algorithms, suggests that not does only the trajectory-GA work, but that it works well. The trajectory-GA optimized trajectories efficiently, and more importantly, reliably. One has to bear in mind that the trajectory-GA itself was not optimized, and was developed as a general trajectory planning optimizer to investigate conceptual methodologies.

The extent of multi-modality imbedded in the trajectory space can be directly linked to the number of arm-configurations that compose a trajectory. Due to technical limitations, that number was kept to a minimum, and therefore, the GA was prevented from showing its full strength in comparison to random search and hill-climbing algorithms.

Chapter 8
LAMARCKISM
AND SUB-GOAL REWARD

Applying sub-goal reward schemes improves our means of learning which sequence of operations leads to an end goal. In addition to the straightforward simplification of the solution space, the increased robustness due to learning in stages is substantial. Sub-goal reward schemes have mainly been applied through sophisticated background operators which attempted to realize intermediate goals explicitly. By seeking explicit knowledge, sub-goal reward schemes often prove to be difficult to construct and too specific to a particular problem domain, a characteristic which reduces their utility. Sub-goal reward schemes can be applied implicitly by incorporating Lamarckian reproduction operators. Lamarckian operators use information that has been acquired through phenotypical adaptation, information which is not normally coded in the genotype (for this reason, Lamarckian information is usually referred to as learned knowledge). Though Lamarckism is not seriously considered as a unified theory that explains the adaptive nature of evolution, it does provide a useful adaptive tool for artificial applications. For such use, many of the disadvantages involved in adopting a Lamarckian framework for evolution do not apply.

This chapter describes Lamarckism in artificial systems and defines a Lamarckian fitness function which realizes sub-goal rewards implicitly in the trajectory-GA environment. When the performance of the trajectory-GA with the Lamarckian operator is compared with the performance of the GA without it, it shows considerable improvements.

Lamarckism

A sub-goal reward is a central issue in all disciplines of learning, both natural and artificial. Any task which consists of several sequential steps leading to a final outcome, raises the issue of sub-goal reward. For if a reward is given only when the end goal is reached, then the longer the sequences are, the smaller the chance of learning how to reach the end goal [Wilson, 1987]. Sub-goal reward means that if an intermediate step has a measurable benefit which is independent of subsequent steps, then this step should be rewarded accordingly and regardless of the success or failure of the final goal. The importance of sub-goal reward lies in the fact that subsequent to the identification of independent sub-goals, the problem can be simplified by the segregation of step sequences into sub-problems. The aspects of sub-goals can be exemplified by the game of chess. Although many good manœuvres can be executed by a player, the game may nevertheless be lost by one grave mistake at the end. Should all game moves be considered as bad moves due to the eventual loss of the game? Realizing that sub-goal reward may speed-up learning, artificial intelligence researchers became interested in how to identify sub-goals and how to properly reward them. The tacit assumption that intelligence is deterministic led to the development of sub-goal deterministic algorithms such as Samuel's checkers program [Samuel, 1959], Winston's system for concept learning [Winston, 1981; Winston, 1984], and reinforcement learning [Jordan and Rosenbaum, 1988].

The Lamarckian theory (better known for its ideas on the inheritance of acquired characteristics) lost its followers when modern research did not support the principal arguments of the theory. Although occasional discoveries bring brief bouts of renaissance to Lamarckism [Berek, *et al.*, 1985; Gorczynski and Steele, 1981], the ability to inherit acquired qualities is generally accepted as non-feasible [Maynard-Smith, 1989], and was explicitly rejected by August Weismann in his central dogma [Weismann, 1904]. Weismann claimed that the organism's cells that experience and respond to phenotype adaptation are not the sex cells that lead to the gametes that form the starting point of the next generation. This is the major reason why many algorithms, simulating or mimicking natural processes, have ignored

Lamarckism and avoided implementing its ideas. The original views of the French biologist J. B. Lamarck (1744-1829) were founded on four basic principles: an inner impulse towards perfection, the ability of an organism to adapt to its environment, the spontaneous creation of life and the ability to inherit acquired characteristics. On the adaptability issue, Lamarck believed that an organ of a species can change and adapt in the course of generations due to constant use or lack of it. The classical example is that of the blacksmith's adapted muscles, which are necessary to his trade. The information needed to develop such muscles does not pass down to the blacksmith's offspring although if it did, it would give them an advantage in the trade. There are many biological flaws in the Lamarckian theory. There is no doubt that the ability to inherit acquired characteristics may improve learning because new generations would not have to start afresh. On the other hand, relying too greatly on ready made knowledge would restrict the ability to learn and adapt which is an essential feature in a changing environment (we shall return to this issue in Chapter 10). Furthermore, if good information can be put back into the gametes, there is also the possibility that non-beneficial or even detrimental information, due to illness and misfortune, will be passed down to successive generations. This argument is the main argument evolutionists hold against the effectiveness and utility of such an evolutionary scheme. However, the flaws apply primarily to the natural world. As was shown in many machine learning applications as well as here, Lamarckism offers several advantages in the adaptation of artificial systems.

In the machine learning sphere, Lamarckism has a different implication. The motivation of artificial systems to use operators which nature 'found' to be disadvantageous, stems from the fact that artificial systems are deterministic in nature, and therefore can utilize the benefits involved in inheriting acquired characteristics without, at the same time, being affected too strongly by their shortcoming. In this way, artificial systems may combine the benefits of the two worlds. Indeed, artificial intelligence applications use many models and operators with impressive benefits, such as the back propagation model [Rumelhart, *et al.*, 1986], and morphology in computer vision [Haralick, *et al.*, 1987]. As far as is known today, these models have no analogy in nature.

Lamarckian Probability for Reproduction

The nature of the trajectory generation, an order dependent process, makes sub-goal reward a plausible strategy. Sub-goal reward is a relatively simple application in the trajectory-GA due to the order dependent quality of the trajectory representation used. It is relatively easy to decompose the trajectory fitness according to its arm-configurations. In decomposition we intend to segregate the total fitness into partial values corresponding to the arm-configurations. Having our fitness function in mind (accumulated deviation from the desired path) we shall attempt to assign the proportional deviation contribution of each of the arm-configurations as a means of measuring its partial fitness to the trajectory. It should be stressed that fitness decomposition does not necessarily imply that it is a correct sub-goal measure. It simply states that there is a decomposition which is somewhat relevant. The partial fitness, $(d_t)_i$, of an arm-configuration (Figs. 49a and 49b) is calculated in the following way:

$$(d_t)_i = \int_{\text{arm-configuration } i-1}^{\text{arm-configuration } i} \left|\text{deviation}\right|, \qquad i = 1, 2, \ldots, l. \qquad (13)$$

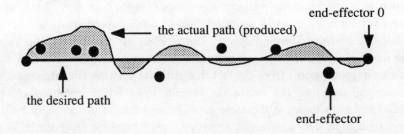

Fig. 49a *- The desired path, the produced path, and the deviation between the two (shaded area). The end-effector positions of the given trajectory are marked for reference.*

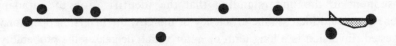

Fig. 49b - *The error contribution attributed to the first arm-configuration (shaded area).*

If the goal function aims at minimizing the accumulated deviation from the desired path, then it is possible to estimate the proportional deviation each displacement vector contributed to the total deviation (Fig. 49). However, and greatly dependent on the amount of nonlinearity incorporated in the representation structure, sub-goal reward is not an accurate or direct measure of the real value of the arm-configuration (the nonlinearity issue was briefly discussed in Chapter 2 and shall be further analyzed in Chapter 9). What we have considered as a sub-goal reward scheme is just one possible scheme. In the robot trajectory application, performance is very much dependent on the relation between adjacent arm-configurations. An unharmonious displacement (discussed in Chapter 5 and illustrated in Fig. 26) between two arm-configurations, which results in a relatively large deviation, is an indication that the two are not co-adapted. If a sequence of arm-configurations has a relatively low deviation contribution, then this sequence is relatively more co-adapted than other sequences in the trajectory (Fig. 50 and 51). Such estimation of sub-trajectory deviation is performed during the evaluation phase.

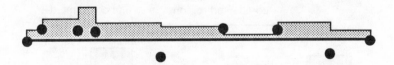

Fig. 50 - *Normalized error distribution segregated according to the arm-configurations.*

This kind of sub-goal information can assist the reproduction operators to predict where it is more promising to operate, and where to concentrate their search activities. It is true that the sub-goal fitness estimate is just an estimate, and may be misleading. We shall justify the expected advantage in using such directed information with the schema theorem. The schema theorem advocates that it is advantageous to crossover on the borderline between co-adapted parameters in order to minimize the disruptive effect crossover has.

Consequent to the understanding that the identification of co-adapted parameters would increase the efficiency of the GA, the inversion operator was suggested. Inversion is a long term operator which decreases the probability of separating co-adapted parameters through crossover [Goldberg, 1987a; Goldberg, 1987b; Holland, 1975]. Although it is impossible to apply the inversion operator in a straightforward way for order dependent strings due to the order quality, it is possible to regain its effect by altering the probability of selecting cross sites according to the deviation distribution along the string [Schaffer, 1987]. If the crossover probability varies according to phenotypic performance, then this is in essence a Lamarckian process. This kind of Lamarckian behaviour has a significant effect on the efficiency and stability of the trajectory-GA (and other GAs [Schaffer, 1984; Schaffer, 1985b]).

Directing the crossover operator probabilistically to operate more frequently at loci of large local deviation is a straightforward sub-goal reward application. But how is it associated with the Lamarckian theory? If the genetic material that is passed on to subsequent generations is directed by the fitness of the parent generation, then this is an information flow from the environment into the reproduction process, a flow direction Weismann's central dogma excludes. Similarly, if a string investigates its constituting information (the displacement vectors in the trajectory application), and as a result of that information alters the course of crossover (determining the resulting offspring trajectory), then acquired information has travelled back into the string.

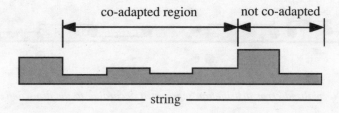

Fig. 51 - *The Lamarckian distributed probability for all the reproduction operators.*

In a fashion similar to varying probability for selecting cross sites, replacing the uniformly distributed probability for mutation, addition, and deletion with a Lamarckian probability will have a significant effect on the efficiency of the mutation, deletion, and addition operators. Using Lamarckian probability, these operators preserve the benefits of introducing a new genetic material, but diminish the counter-productive effects such material may have. By directing mutation efforts at loci that are not co-adapted, or have a low fitness contribution to the global fitness of the trajectory, the mutation, deletion and addition operators still introduce new information to the population, but do not tend to destroy good information already discovered. The Lamarckian probability enables these operators to be more controlled, and to introduce some qualities of a hill-climbing algorithm.

Results and Discussion

To analyze the effect of Lamarckian probability, the trajectory-GA was applied twice to a set of trials. Once the reproduction was based on uniform probability and once on a Lamarckian probability. The comparative study between the two experiments is summarized in Table 4 and Figs. 52 through 57, and covers a total of 20,000 trials. It should be remembered that the Lamarckian probability represents a subtle change in the way the GA operates. Instead of choosing an arm-configuration as a cross site with equal probability, the arm-configurations are chosen probabilistically according to their respective local error. Nevertheless, under the Lamarckian probability regime any arm-configuration can be selected as a cross site, but with a varying probability. What are the gains one should expect from thus changing the probability on which the   reproduction operators base their operative

	1	2	3	4	5	Average	SD
uniform probabilities	1.1090	0.5591	1.1532	1.3101	1.2322	1.0727	0.2972
Lamarckian probabilities	0.7920	0.4707	0.7668	0.5246	0.5087	0.6126	0.1538

Table 4 - *Statistical comparison between the performance of a trajectory-GA with a uniform probability and that with Lamarckian probability.*

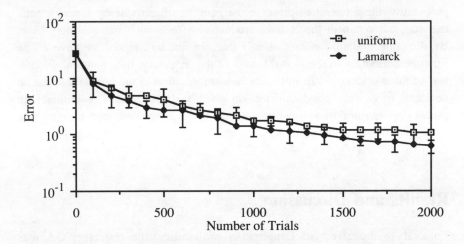

Fig. 52 - *Statistical summary of five pairs of experiments comparing the performance of a GA with uniform probabilities for reproduction (solid squares) and Lamarckian probabilities (hollow diamonds). This statistic is based on 20,000 trials.*

decisions? In general, Lamarckian probability (as defined here) has a beneficial and noticeable effect over performance provided the solution space exhibits two features:

(1) Correlation between the local error distribution and true performance is fairly good (low order of nonlinearity in the representation).

(2) Error distribution is significantly non-uniform.

The meaning of the first condition was discussed earlier. The second condition requires a large diversity of error distribution along the trajectories because otherwise, the Lamarckian probability degenerates into a uniform probability crossover. Resulting from the above two conditions, the GA using the Lamarckian probability should exhibit a significantly different performance only at search stages where the diversity of co-adapted arm-configuration sequences is sufficiently large.

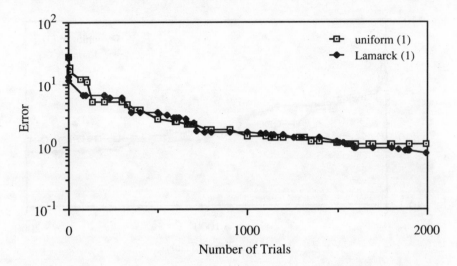

Fig. 53 - *Comparison between the effect of Lamarckian and uniform reproduction probabilities on search progress (experiment 1).*

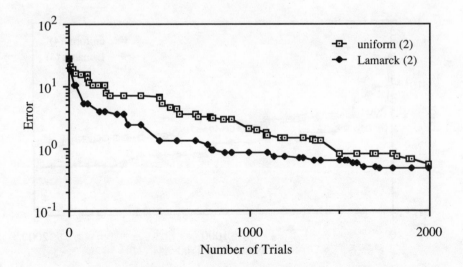

Fig. 54 - *Comparison between the effect of Lamarckian and uniform reproduction probabilities on search progress (experiment 2).*

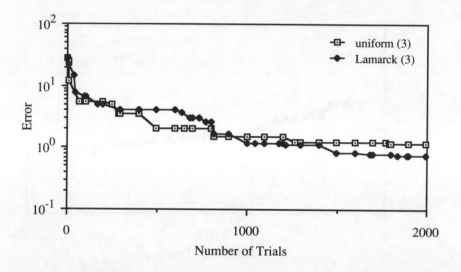

Fig. 55 - *Comparison between the effect of Lamarckian and uniform reproduction probabilities on search progress (experiment 3).*

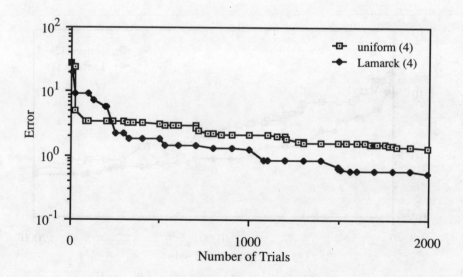

Fig. 56 - *Comparison between the effect of Lamarckian and uniform reproduction probabilities on search progress (experiment 4).*

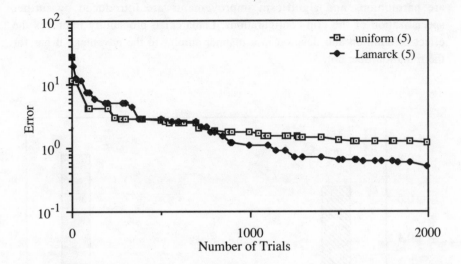

Fig. 57 - *Comparison between the effect of Lamarckian and uniform reproduction probabilities on search progress (experiment 5).*

A large number of generations is required before the effect of the Lamarckian probability distribution can be clearly observed. The limited number of overall trials, generated at each experiment, reduces the effectiveness of the Lamarckian probability over uniform probability. The dynamic resolution mechanism reduces the trajectory diversity and therefore substantially reduces the effectiveness of the Lamarckian probability. The last factor that reduces the efficiency of the Lamarckian probability over the uniform case, is the relatively small number of arm-configurations used. The smaller the number of arm-configurations, the fewer the plausible number of co-adapted sub-sequences. The fewer co-adapted sequences a trajectory can include, the closer the Lamarckian distribution is to a uniform one, an undesirable situation when applying a Lamarckian probability.

The error distribution along the trajectory is also influenced by the remaining reproduction operators, namely mutation, deletion and addition. These operators, when Lamarckian probability is applied, become significant at advanced stages of the search. The main advantage expected from a Lamarckian mutation is the concentration of mutation activities at loci of relatively poor fitness, and the consequent reduction of the disruption. This effect is more noticeable at advanced stages of the search, when the trajectories are harmonious and significant improvements are introduced by further specialization of the arm-configurations. Lamarckian probability improves the effect of addition and deletion in a manner similar to the advantage it has for the mutation operator.

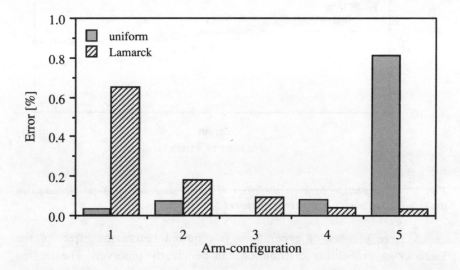

Fig. 58 - *A comparison between the typical error distribution along the best trajectory in the case of the uniform and Lamarckian probability reproduction operator.*

There is an interesting phenomenon concerning the error distribution along trajectories. When one plots the error distribution along the best trajectory, one finds that trajectories have a characteristic deviation distribution which is different in the two probability strategies (Fig. 58). At the end of the optimization process under the uniform distribution, most of the error of a trajectory is concentrated at its last arm-configuration. When the Lamarckian probability is used, most of the error of a trajectory is concentrated at its first arm-configuration (crossed bars in Fig. 58). The characteristically different error distribution of the two strategies supports the argument that Lamarckian mutations tend to focus on poor fitness loci and thus introduce new genetic material where it is most needed. But why is the error distribution under the Lamarckian probability not significantly more uniform than what it is? The error contribution of the last arm-configuration is mainly dependent upon the last arm-configuration itself, and therefore, mutation activity has a pronounced influence over the error contribution of that arm-configuration. Directing mutation efforts to the last arm-configuration has the effect of increasing the probability of improvement at that locus. All intermediate arm-configurations are mainly dependent on co-adaptation to adjacent arm-configurations and therefore, Lamarckian mutations have a lesser effect. Because the initial arm position is fixed for all trajectories, and the first arm-configuration has relatively small arm flexibility (Eqs. 9 and 10), it is difficult to optimize the first arm-configuration through mutations. Fig. 58 presents a comparison of a typical error distribution along the best trajectories under the two probability distributions.

The interpretation of Fig. 58 requires some care. For reasons discussed above, the full effect of the Lamarckian operators cannot be exhibited in the experimental setup used here. However, the final arm-configuration is a special case where the Lamarckian probability has a prominent effect. This is because of the proportionately greater error contribution of the final arm-configuration, an artifact of the fitness function (Eq. 10). Because the last arm-configuration does not have to consider a subsequent displacement to another arm-configuration, the co-adaptation element becomes less important. The main error component of the last arm-configuration is due to the distance of its end-effector from the end of the desired path, which is independent of other arm-configurations. Thus, whilst the differences in the partial errors for arm-configurations 1-4 do not show significant variance, the improvement of the partial error at position 5 is readily attributed to Lamarckian mutations.

Summary

This chapter explored the possibility of identifying sub-goals without compromising the simplicity and applicability of a GA. The sub-goals were discovered implicitly by a Lamarckian error distribution. The applicability of such sub-goal reward strategy is possible due to the relatively low nonlinearity imbedded in the trajectory space. The reward scheme is called sub-goal reward because it is capable of identifying co-adapted intra-trajectory elements and therefore of increasing the likelihood of crossover occurring outside these co-adapted groups. The reward scheme is called Lamarckian because it is based on phenotypic information which is put into the genotype.

The application of the Lamarckian fitness function emerged naturally from the robot trajectory generation problem. In spite of the unfavourable test conditions, the Lamarckian functions exhibited characteristic improvements in the form of stability and efficiency. More work is necessary to quantify these improvements. Such investigation should follow two avenues: one, to widen the experimental suit, namely, to increase the number of generations and to increase the length of the chromosomes, and the other, to apply the Lamarckian function to other problem domains. The main strength of the Lamarckian distribution lies in the fact that it can be applied carelessly, meaning that the Lamarckian distribution has no effect when genes are completely not co-adapted or co-adapted extensively. In either case, there is no reduction in the effectiveness relative to the uniform probability. The main benefit the Lamarckian probability has to offer is the implicit identification of co-adapted string elements, and the concentration of mutation resources at loci which are not co-adapted.

Chapter 9

EPISTASIS IN GENETIC ALGORITHMS

We return in this chapter to the issue of the suitability of a given representation to a GA, an issue we have briefly discussed at the end of Chapter 2. Throughout the book we have stressed the major role the choice of a code has in affecting the efficiency of a GA, and that the GA optimizes the code of the problem to be solved and not the problem directly. This is a fundamental aspect of GAs and suggests that we have to investigate the representation more closely and preferably with systematic analysis in order to assess its suitability to a GA processing.

The number of works which have addressed the suitability issue stand in contrast to its fundamental importance. Much of the past GA research involved a heuristic analysis of the interaction between functions and codings. Hitherto few have suggested a systematic analysis. The problems of calculating schema average fitness resulted in a situation where often apart from the 'awareness' of the fact that the coding is an illusive domain, little was done to improve the understanding of the coding-function relationships. The fact that researchers were not too enthusiastic about the tedious computations involved in calculating schema average fitness is all too understandable when there were no systematic alternatives. However, as early as 1980 Bethke [Bethke, 1980] suggested an efficient method for calculating schema average fitness for the common GA coding format (the fixed length binary representation). Bethke's use of *Walsh functions* to calculate average schema fitness offers an efficient method which provides much needed insight into the workings of the schema processing and into what makes some functions hard for a GA application, yet others simple. Goldberg returned to Bethke's original work and produced a series of manuscripts in which he re-introduced the Walsh function analysis in a clear and digestible format [Bridges and Goldberg, 1989; Goldberg, 1988; Goldberg, 1989a].

The Walsh function analysis is an algebraic method with which a binary fixed length representation is projected onto an auxiliary parameter space where the parameters are the Walsh coefficients. One of the nice attributes of the Walsh coefficients is that they are orthogonal, thus providing a linear decomposition of the schema fitness space. Bethke's ideas received little attention probably due to a deceptive impression that one needs to get involved in heavy mathematical calculations to apply them. The Walsh functions exhibit considerable capacity to analyze the common and widely used fixed length binary representation. Whether that promise will be justified or not depends on further research, but mainly on their greater use in future GA research. Notwithstanding their utility, one also has to be aware of some limiting aspects of the Walsh functions. The limitations are that one has to calculate the Walsh coefficients for each application, and there are a great many of them (there are 2^l coefficients for codes of length l). Furthermore, the Walsh analysis is restricted to representations which are fixed in length and fully determined, a coding philosophy we argued to be restrictive. This issue will be stressed again in Chapter 10.

The somewhat abnormal representation structure we used for the trajectory generation prevented the application of the Walsh functions to our domain. Furthermore, extensions to the theory of the Walsh analysis, to make it more applicable to less rigid representation formats (like trajectory coding), are not forthcoming. There are possibly other methods besides the Walsh functions by means of which the suitability of a representation can be assessed. We propose in this chapter a different approach to the suitability and applicability aspect. The method we shall present is based on the biological term epistasis and on what epistasis comes to designate in population genetics (see preliminary discussion in Chapter 2).

The Linear Assumption

Another aspect of GAs and their coding paradigm is that any fitness function can ultimately be reduced to a set of linearly independent partial fitness functions [Goldberg, 1988] so that for any string j it is possible to write its fitness as the following sum:

$$v(S^j) = \sum_{i=1}^{2^l} f(S^j)\delta_{ij} , \qquad \delta_{ij} = \begin{cases} 1, & \text{if } i = j \\ 0, & \text{if } i \neq j \end{cases} . \tag{14}$$

In other words, theoretically a fitness space can always be reduced into a table of fitness values for each of the phenotypes. This approach is adopted here, but in a different way. Instead of decomposing the fitness space according to strings as Eq. (14) implies, the fitness space is decomposed according to the coding elements (genes' value or alleles). Assuming such a decomposition is possible, the fitness of any string j may be calculated by summing the value of its genes:

$$A(S^j) = \sum_{i=1}^{2l} A(S_i^j) . \tag{15}$$

This means that instead of the 2^l fitness values required to compute any fitness according to Eq. (14), only $2l$ values are needed when considering Eq. (15). Furthermore, the discussion here focuses on the relation between the two values, and its use as a suitability criterion for GAs efficiency.

The objective for applying the above linear decomposition is to develop a method for the prediction of the amount of nonlinearity (in terms of gene interaction) embedded in a given representation. To this end, fitness has to be associated with the representation elements. If a linear decomposition proves to be inaccurate, then it implies that the representation incorporates nonlinearities. Quantifying the amount of nonlinearity will provide an estimate for the suitability of a given representation to a GA processing. From a GA perspective, a coding format in which the effect of any individual parameter on the total fitness is independent of other parameters suggests that there is little co-adaptation.

On the other hand, a high degree of nonlinearity indicates that above average schemata are too long. The whole GA ideology is based on the assumption – that one can only say something about the whole by knowing its parts. Neither the schema theorem nor population genetics indicate exactly how much of the whole the parts should indicate.

It is possible to detect nonlinearity by measuring the discrepancy between the real fitness and the recomposed fitness according to Eq. (15). The arguments for estimating the degree of nonlinearity of a coding function by estimating the applicability of the linear assumption are sound and are implicitly founded in the schema theorem. What is less clear is how these ideas can be formulated in a practical way. This issue is discussed in the next section.

Basic Elements of Epistasis

It was already emphasized above that the effect of epistasis lies in the ability to predict the value of a whole from the value of its parts. One possible method of calculating epistasis is based on the linear assumption and is loosely connected to Fisher's theorem (see [Crow and Kimura, 1970] for a detailed discussion on Fisher's theorem). The following definitions are adopted for the preliminary analysis:

A string, S, is composed from l elements s_i (without loss of generality, l is fixed),

$$S = (s_1, s_2, \ldots, s_l) . \tag{16}$$

Without loss of generality, only a binary alphabet is considered. The allele of the ith gene in a string is denoted by

$$s_i = a \qquad a \in \{0, 1\} , \qquad i = 1, 2, \ldots, l . \tag{17}$$

The *Grand Population*, Γ, is the set of all possible strings of length l,

$$\Gamma = \prod_{i=1}^{l} \{0, 1\} . \tag{18}$$

Let *Pop* denote a sample from Γ where the sample is selected uniformly and with replacement. The size of a sample *Pop* is

$$N = |Pop| . \tag{19}$$

The fitness of a string is given by

$$v(S) = \text{fitness} \tag{20}$$

where v is a 'black box' function. The average fitness value of the sample Pop is

$$\bar{V} = \frac{1}{N} \sum_{S \in Pop} v(S) . \tag{21}$$

The excess fitness value of a string is denoted by

$$X(S) = v(S) - \bar{V} . \tag{22}$$

The number of string instances in Pop which match $s_i = a$ is denoted by $N_i(a)$. The average allele value is denoted as

$$A_i(a) = \frac{1}{N_i(a)} \sum_{S \in Pop_{s_i=a}} v(S) \tag{23}$$

where $Pop_{s_i=a}$ is the set of all strings in Pop having the allele a in their ith position. The weight of s_i is

$$\Delta_i = |A_i(1) - A_i(0)| . \tag{24}$$

The excess allele value is defined by

$$X_i(a) = A_i(a) - \bar{V} , \tag{25}$$

the excess genic value is

$$X(A_i) = \sum_{i=1}^{l} X_i(a) \tag{26}$$

and the genic value of a string S – the predicted string value – is defined as

$$A(S) = X(A_i) + \bar{V} . \tag{27}$$

Thus, the difference $\varepsilon(S) = v(S) - A(S)$ might reasonably be supposed to be a measure of epistasis of a string S.

Consequently, an epistasis measure for the Grand Population and hence for the representation, is termed the *epistasis variance* and is defined as

$$\sigma_\varepsilon^2 = \frac{1}{N_\Gamma} \sum_{S \in \Gamma} [v(S) - A(S)]^2 \tag{28}$$

where the implicit $A_i(a)$ are computed over the Grand Population (note that this definition does not follow the common definition of variance as it involves elements from two different sets). This measure can be estimated from the corresponding expression

$$\sigma_{Pop}^2 = \frac{1}{N} \sum_{S \in Pop} [v(S) - A(S)]^2 . \tag{29}$$

However, since the computation of $A_i(a)$ is determined by the sample population, this statistic is subject to sampling error (parasitic epistasis), but as yet, confidence measures for the estimate are unavailable. This would require an investigation of the distribution of

$$\sigma_\Gamma^2 - \sigma_{Pop}^2 .$$

The above definitions (summarized in Table 5) provide a method for estimating the epistatic variance for a Grand Population – the base epistasis – from a sample population. The distinction between base epistasis and parasitic epistasis is very important because the effect of the latter is often of equal or even higher order of magnitude. This will be demonstrated further subsequently.

The fitness variance is denoted as

$$\sigma_v^2 = \frac{1}{N} \sum_{S \in Pop} (X(S))^2 \tag{30}$$

and the genic variance is denoted as

$$\sigma_A^2 = \frac{1}{N} \sum_{S \in Pop} (X(A_i))^2 . \tag{31}$$

The difference between the fitness variance and the genic variance is important (though not intuitive) for estimating to what extent the sample departs from the Grand Population and is denoted as

$$\sigma^2_{v-A} = \sigma^2_v - \sigma^2_A \ .$$ (32)

Symbol	Term
S	String
$v(S)$	Fitness
$X(S)$	Excess fitness value
a	Allele
$A_i(a)$	Allele value of a
$X_i(a)$	Excess allele value
$X(A_i)$	Excess genic value
$A(S)$	Genic value
$\varepsilon(S)$	Epistasis value
σ^2_v	Fitness variance
σ^2_A	Genic variance
σ^2_ε	Epistasis variance

Table 5 - *Summary of the symbols and their definitions in the epistasis discussion.*

Calculating Epistasis for Two Illustrative Functions

We shall demonstrate the use of the formulae developed in the previous section. We shall calculate the epistasis variance for two illustrative functions. To demonstrate the simplicity by which nonlinearity is detected when considering epistasis, we choose two epistatically extreme functions, a zero epistasis function, and a function of maximum epistasis. The two functions are the algebraic function *SUM*mation of zero epistasis (Fig. 59),

$$SUM = 2.33 \sum_{i=1}^{3} s_i \ , \qquad s_i = \{0, 1\} \ ,$$ (33)

and the logical function *AND* of a maximum epistasis (Fig. 60),

$$AND = 28 \prod_{i=1}^{3} s_i , \qquad s_i = \{0, 1\} . \tag{34}$$

Our first analysis shall use grand populations and thus addresses the issue of base epistasis. The functions are scaled so that the average genotype value is equal in both. This is necessary to achieve comparability between the epistasis variance in the absence of standard normalizing procedures. Tables 6 and 7 summarize the values of the various terms involved in calculating the epistasis variance. The procedure is the following: one first calculates the average alleles value (Eq. 23), and their corresponding excess value (Eq. 25). These

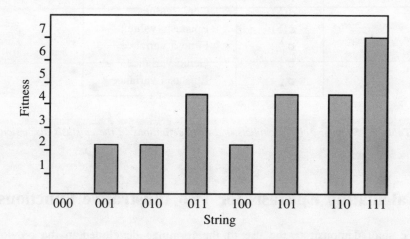

Fig. 59 - Genotypes and fitnesses of the SUM function.

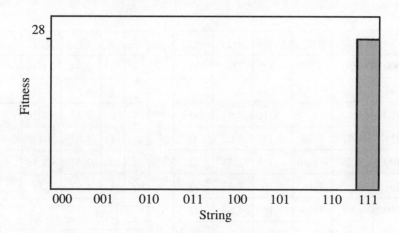

Fig. 60 *- Genotypes and fitnesses of the AND function.*

values are summarized in the right box in both tables. It is now possible to calculate the excess genic value (Eq. 26 and third column from the right in the left box), and the genic value for each genotype (Eq. 27 and second column from the right in the left box). Now, subtracting the genic value from the genotype value gives the epistatic value for that genotype (rightmost column in the left box). At this stage and without calculating the variance one can immediately see the traces of epistasis. Each of the genotypes for the *SUM* function can be calculated accurately from the statistic, but not those of the *AND* function which is nonlinear.

We calculate the fitness variance (Eq. 30), genic variance (Eq. 31), epistasis variance (Eq. 28), and the statistic for the departure from the grand population (Eq. 32) all of which are summarized in the lower right box. The epistasis variance for the linear function *SUM* is zero while the epistasis variance for the nonlinear function *AND* is not zero (we shall later address the issue of scaling epistasis variance with a normalized measure).

S	$v(S)$	$X(S)$	$X(A_i)$	$A(S)$	$\varepsilon(S)$
000	0	−3.5	−3.5	0	0
001	2.33	−1.16	−1.16	2.33	0
010	2.33	−1.16	−1.16	2.33	0
011	4.67	1.16	1.16	4.67	0
100	2.33	−1.16	−1.16	2.33	0
101	4.67	1.16	1.16	4.67	0
110	4.67	1.16	1.16	4.67	0
111	7	3.5	3.5	7	0

i	a	$A_i(a)$	$X_i(a)$	Δ_i
1	0	2.33	−1.16	2.33
	1	4.67	1.16	
2	0	2.33	−1.16	2.33
	1	4.67	1.16	
3	0	2.33	−1.16	2.33
	1	4.67	1.16	

σ_v^2	σ_A^2	σ_ε^2	$\sigma_v^2 - \sigma_A^2$
4.08	4.08	0	0

Table 6 - *Calculating the epistasis variance for the SUM function*

S	$v(S)$	$X(S)$	$X(A_i)$	$A(S)$	$\varepsilon(S)$
000	0	−3.5	−10.5	−7	7
001	0	−3.5	−3.5	0	0
010	0	−3.5	−3.5	0	0
011	0	−3.5	3.5	7	−7
100	0	−3.5	−3.5	0	0
101	0	−3.5	3.5	7	−7
110	0	−3.5	3.5	7	−7
111	28	24.5	10.5	14	14

i	a_i	$A_i(a)$	$X_i(a)$	Δ_i
1	0	0	−3.5	7
	1	7	3.5	
2	0	0	−3.5	7
	1	7	3.5	
3	0	0	−3.5	7
	1	7	3.5	

σ_v^2	σ_A^2	σ_ε^2	$\sigma_v^2 - \sigma_A^2$
85.75	36.75	49	49

Table 7 - *Calculating the epistasis variance for the AND function.*

By analyzing the epistasis variance of the *SUM* and *AND* functions, one can observe the strength of the linear assumption. The *SUM* function can be accurately composed of the decomposed $A_i(a_i)$ values, while the re-composition of the *AND* function reveals a large epistatic variance.

S	$v(S)$	$X(S)$	$X(A_i)$	$A(S)$	$\varepsilon(S)$
000	0	−3.5	−4.29	−0.79	0.79
001	2.33	−1.16	−1.95	1.55	0.78
010	2.33	−1.16	−1.95	1.55	0.78
101	4.67	1.16	1.95	5.45	−0.78
110	4.67	1.16	1.95	5.45	−0.78
111	7	3.5	4.29	7.79	−0.79

i	a_i	$A_i(a)$	$X_i(a)$	Δ_i
1	0	1.55	−1.95	3.9
	1	5.45	1.95	
2	0	2.33	−1.17	2.34
	1	4.67	1.17	
3	0	2.33	−1.17	2.34
	1	4.67	1.17	

σ_v^2	σ_A^2	σ_ε^2	$\sigma_v^2 - \sigma_A^2$
4.98	8.67	0.61	−3.69

Table 8 - *A moderate partial population shows a parasitic epistasis variance. The shaded strings are the ones not included in the statistics.*

S	$v(S)$	$X(S)$	$X(A_i)$	$A(S)$	$\varepsilon(S)$
000	0	−3.11	−7.38	−4.27	4.27
001	2.33	−0.78	−2.32	0.79	1.54
111	7	3.89	9.34	12.45	−5.40

i	a_i	$A_i(a)$	$X_i(a)$	Δ_i
1	0	1.17	−1.94	5.83
	1	7	3.89	
2	0	1.17	−1.94	5.83
	1	7	3.89	
3	0	0	−3.5	5.06
	1	4.67	1.56	

σ_v^2	σ_A^2	σ_ε^2	$\sigma_v^2 - \sigma_A^2$
8.47	49.03	16.59	−40.56

Table 9 - *A smaller sample exhibits greater parasitic epistasis variance.*

Partial Populations and Sampling Noise

Since the population size in all practical GA applications is only a minuscule portion of the genotype pool, it is important to understand and quantify the effect a sample population has over the epistasis variance. The sampling error and the induced parasitic epistasis are analyzed for the *SUM* function. This demonstrative example suggests that the sampling error has an overpowering effect over the epistasis variance. A similar analysis for the *AND* function and other functions suggests that the parasitic epistasis is primarily a function of the size of the sample investigated. It can also be shown mathematically that the difference between Eq. 28 and Eq. 32 is a non-zero sum for all cases where the population is not the grand population.

Using the *SUM* function to illustrate this matter we observe that when the grand population is used (Table 6), no epistasis is measured. This however changes dramatically when only a sampled population is used (Tables 8 and 9), and a parasitic epistasis is introduced.

Summary and Conclusions

We have discussed in this chapter some basic problems regarding coding formats and GAs. It was suggested that much of future success in GAs research is dependent on the development of systematic methods by which a given coding of a problem domain can be optimized so as to maximize the efficiency of a GA processing of that particular application. It was also suggested that by understanding what makes problems simple or hard for a GA, one shall understand the workings of GAs better. In spite of its importance, analyzing representations precisely is not common due to the tedious computation involved. There is a novel approach to analyzing coding-function relationships proposed by Bethke which considered the use of Walsh functions to create an auxiliary mapping space of the fitness space, an auxiliary space by which the average fitness of schemata is readily calculated. As an alternative method to simplify the coding enigma, and in an attempt to overcome some limitations of the Walsh function analysis, the epistasis variance analysis was suggested. Epistasis variance is a more flexible method by which the degree of nonlinearity imbedded in a coding format can be measured. The measure of nonlinearity is indirect and cannot differentiate between different orders of nonlinearity, but in certain aspects it is very simple and intuitive. Measuring epistasis is based on the linear assumption that the coding parameters are linearly independent with respect to the fitness function. Assuming linear independence, one simply decomposes fitness according to average parameters' value, and attempt to compose the original fitnesses by these values. The accuracy of this assumption, or more precisely the variance of this method of calculating fitnesses over a sample population, is an estimate of the total amount of nonlinearity imbedded in the coding. The method proposed is not conclusive, but certain underlying coding-function aspects can be learned from it.

Calculating the epistasis variance in grand populations of two illustrative functions (and for other functions the results of which were not presented here) reveals the following:

(1) The epistasis as defined in this work detects <u>all</u> orders of nonlinearity.
(2) The epistasis variance cannot be scaled or parameterized with the tools presented here, but can be compared qualitatively.

The epistasis variance is usually determined by a sampled population. Analyzing the epistasis variance for different samples reveals an extensive sampling error resulting from such approximation. Confidence measures for the extent of this approximation were not presented. Epistasis is defined in terms of two elements: the base epistasis and parasitic epistasis. It is important to distinguish between the two because only the base epistasis is relevant to the issue of suitability between a coding and the GA. The parasitic epistasis can help in determining the optimal population size. Noting the remaining difficulties in applying epistasis measurements, the epistasis variance does provide two important lessons which should be stressed:

(1) It is possible to measure the extent of nonlinearity without knowing anything about the fitness function. The analysis holds for all representations where the elements of the representation can be identified.
(2) The common use of epistasis in the context of nonlinearity is misleading because it combines two nonlinear properties: a base epistasis and a parasitic epistasis.

Chapter 10
AN ADAPTATION ANOMALY

The purpose of this chapter is twofold: To analyze an adaptation anomaly observed in the trajectory-GA, and to propose an explanation for this unusual adaptive behaviour by drawing an analogy to some elementary adaptive mechanisms in nature. Adaptation as used in this chapter means the ability to adapt to arbitrary conditions. This definition encompasses adaptive systems which optimize their ability to adapt while adapting to given conditions. The optimization of the ability to adapt in many natural systems involves the initial relative simplification of the system, followed by an increase in sophistication. Such bottle-neck behaviour seems to be a strategy of many natural systems on both the macro- and the micro-levels. Its abundance in nature as well as its evolutionary persistence suggests that, in terms of fitness, the bottle-neck behaviour is an evolutionary stable strategy of high fitness.

The bottle-neck phenomena are presented here as anomalous simply because much of the work on evolutionary systems, and especially long term evolution, regards adaptation as a monotonous process during which simple organisms evolve into more sophisticated organisms by improving their information structure. This view of evolution is correct, to a large extent, in long term adaptation and evolution, but not in short term adaptation within an organism's lifetime or within a few generations. In contrast to large scale evolutionary processes, where complex and more intricate species evolve from

131

simpler forms of life, many short term adaptive processes in higher organisms choose to start from a state of relatively high complexity and redundancy (over-specification). The state of relative redundancy evolves into a state of relative simplification as a result of competition and selection, which in turn evolves again into a state of relatively high complexity and sophistication. This process which is initially characterized by a state of disorder and relatively high complexity, which decreases and later increases again as the organism adapts, is termed here as *adaptation anomaly*.

The anomaly is presented as it is observed in the trajectory-GA. Its anomalous aspects will be explained in respect to the trajectory generation domain. Thereafter, few elementary natural systems which exhibit a similar adaptive behaviour to that of the trajectory-GA, will be reviewed. Finally, a more formal definition of the anomaly, and of the information content of an adaptive system will be suggested. These definitions will provide an interpretation of the anomaly which will offer a qualitative explanation for the bottle-neck type processes in terms of fitness considerations.

The Anomaly

We are familiar with the robot model, where the number of arm-configurations defining a trajectory is dynamic. The dynamics of the number of arm-configurations defining a trajectory as a function of the optimization process can now be examined. In other words, the average number of trajectory specifications as a function of the evolving population of trajectories, is examined. In general, if the number of arm-configurations defining a trajectory has no substantial effect on the performance of trajectories, one would expect that when starting from almost any distribution of the number of arm-configurations, the average number of arm-configurations in the whole population does not change during the optimization process (at least not drastically). Individual trajectories may lose a few arm-configurations, but others would gain, and on average it would balance out. This expectation should still hold, even when the deletion and addition operators are introduced because of the equal and small frequency in which these operators are used. Furthermore, even if the number of arm-configurations has an effect on performance, a monotonous behaviour of the average number of arm-configurations should be expected. The monotonous growth or decline in the number of arm-configurations can be expected because neither the algorithm nor the fitness function changes during the optimization cycle. Therefore, since the environment is fixed, the effect on performance of changing the number of arm-configurations should be consistent, either improving or impairing it. If having more arm-configurations introduces some novelty to the trajectory, and improves its performance, then the average number of arm-configurations in the population should monotonously increase until some saturation level has been reached. The same arguments hold for the opposite case. If having fewer arm-configurations improves performance, then the average number of arm-configurations in the population should monotonously decline to some threshold level. Whatever effect the number of arm-configurations has on trajectory performance, it would not be logical to expect that an increase will follow a decrease in the average number while fitness is monotonously improving.

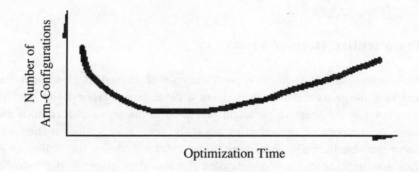

Fig. 61 - *A qualitative description of the adaptation anomaly (or bottle-neck) in the trajectory-GA.*

Yet, in contrast to all well-founded arguments presented above, the trajectory-GA behaves differently. The average number of arm-configurations characteristically declines during initial stages of the optimization, but at some stage stabilizes and then starts to increase (see Fig. 61 for a qualitative behaviour of the bottle-neck phenomenon). We have termed adaptation anomaly the phenomenon where the average number of arm-configurations is being reduced due to selection pressure favouring short trajectories, but at some later stage, and due to the same selection pressure, starts to reward long trajectories which consist of more arm-configurations.

Adaptation Anomaly in Nature

Before suggesting a few explanations for the anomaly and attempting to draw some general behavioural conclusions, a few similar adaptation phenomena in nature will be analyzed. Three illustrative examples from nature are presented, examples which exhibit the bottle-neck behaviour and cover both micro- and macro-systems. The three examples from nature are by no means the only examples of the bottle-neck behaviour and form only a small collection of similar phenomena. In a sense, and because the natural systems that exhibit the so called 'adaptation anomaly' are so fundamental, this behaviour should probably not be regarded as anomalous. The reason for describing these processes as anomalous is only because they are counter-intuitive. The examples given below are from the following domains: pre-attentive human vision, population genetics, neurogenesis (the development of a nervous system), and the development of a single neural cell.

Pre-attentive Human Vision

Pre-attentive human vision is a good example of anomaly as it is both clear and very illustrative. Although it is not a development process and therefore does not suit the adaptive paradigm presented in the previous section, it does show one aspect of the adaptive anomaly very clearly: information as a parameter should really be viewed as the balance between the quality of the data and its quantity (a more detailed discussion is given in the following section).

Recent research in human vision suggests that vision is composed of two distinctly different processes, pre-attentive and attentive vision. Pre-attentive vision operates in parallel in the sense that it can process visual stimuli in several ways simultaneously, while attentive is mainly a serial process which can only use limited visual stimuli. It is of great theoretical interest to understand what types of visual information are 'considered' by the vision system as elementary independent features which can, and should, be processed in parallel (one can observe the relevancy of the existence of such special visual features to the development of appropriate visual mechanisms). Several works have addressed this issue, but we shall focus only on one characteristic aspect of pre-attentive vision – the ability to detect a line object (Fig. 62). It was reported [Sagi and Julesz, 1987] that the ability to detect an

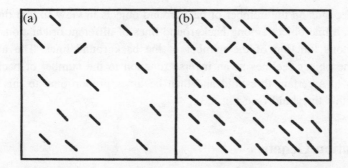

Fig. 62 *- Two illustrations used as stimuli in measuring the response of an observer asked to detect the presence of the line having a different orientation from the others. It was discovered that the response for detection is not monotonous with the number of background lines. (Adapted by permission from "Short-range limitation on detection of feature differences", copyright by Dov Sagi and Bela Julesz, Spatial Vision, Vol. 2, No. 1, p. 40).*

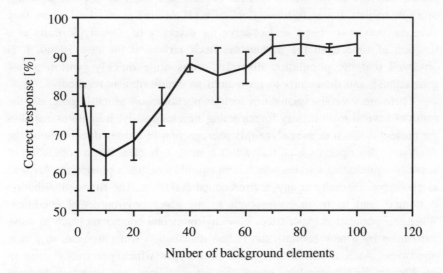

Fig. 63 *- Average response for different elements presented on the retina (horizontal and vertical lines, diagonal lines, and horizontal and diagonal lines) of three observers. The response is initially reduced as a result of an increase in the background elements, but after a certain background density it increases. (Adapted by permission from "Short-range limitation on detection of feature differences", copyright by Dov Sagi and Bela Julesz, Spatial Vision, Vol. 2, No. 1, p. 45).*

object depends on the number of background objects in view so that the ability to detect a line object among background lines of different orientation, is not a monotonous function of the number of the background lines. The ability to detect the object declines as an inverse function to the number of background lines up to a certain threshold and then becomes proportional to this number beyond this threshold (see Fig. 63).

Population Genetics

The term 'bottle-neck' has so far been used to describe the dynamics of a phenomenon which characterizes a specific adaptive behaviour. This term originates from population genetics, where it describes a phenomenon in which a population's diversity during the population's adaptation or specialization is initially reduced but later increases again. This phenomenon is frequently observed when there is a major change in the selective pressure acting on a population, such as the drastic changes experienced when the population settles in a new habitat. Let us follow a flock of birds arriving at an island where they have not previously been, and observe the diversity of phenotypic traits as a function of time from the minute the flock arrives at the new island. It is observed that the population diversity drops quite quickly (within a few generations), and then starts to grow until an equilibrium is reached.

There are several explanations for this phenomenon which is generally the result of several evolutionary forces acting simultaneously. It is not essential in the present discussion to analyze this phenomenon in greater detail beyond the illustrative description which follows. It is intuitively clear that a population of sexually reproducing species which is in equilibrium, has a phenotypic (as well as genotypic) diversity at any environmental condition. The fitness distribution is normal, and its mean corresponds to the given environmental condition. When this population experiences new environmental conditions (such as those introduced by a new habitat), the fitness distribution shifts its peak to a new phenotype. As a result of this shift, phenotypes which previously were in equilibrium find themselves below the survival level and therefore become diminished. These phenotypes will become extinct and the diversity of individuals will consequently be reduced. With time, reproduction creates new phenotypes and the population diversity grows until a new equilibrium is reached [Maynard-Smith, 1989]. Here again, the adaptation anomaly, which in the course of adaptation passes through a bottle-neck, can be observed.

The Development of a Nervous System

The development of a nervous system is a perfect example of the adaptation anomaly. It is adaptation occurring at the level of individual organisms, a lower level than the macro-level of whole populations given in the previous example. After the initiation of the development of any part of the brain, the consequent development follows a characteristic dynamic sequence described by the following chronological stages: the proliferation of cells, the differentiation of the immature neurons, the formation of connections with other neurons, the selective extinction of certain cells and elimination of some of the connections that were initially formed, and finally the stabilization of the remaining cells and their further growth [Cowan, 1979; Kandel, 1979]. This counter-intuitive and implausible nature of development, in which an initial redundancy due to an over-development is reduced by a massive simplification of the system (a large number of cells and connections are eliminated), only later to become relatively complex again, due to growth in the number of cell connections and specialization of the existing cells, is not a result of a capricious process. On the contrary, the simplification and consequently sophistication is closely associated with the improved fitness of the organism. Here yet again the bottle-neck behaviour and the adaptation anomaly can be observed.

The fact that biological systems develop and grow is obviously not the puzzling part of the fundamental development cycle which was just described. Rather, it is the massive simplification of the system which precedes growth which is puzzling. Why would it be advantageous for an organism to develop its nervous system in this seemingly 'fickle' way? On that, Oppenheim comments:

"The death of some cells during normal development is perhaps not an entirely unexpected phenomenon in that a certain amount of genetic and epigenetic imprecision is bound to exist in any biological system, giving rise in some instances to circumstances (either intrinsic and/or extrinsic in origin) that may be lethal to developing cells. When the extent of such cell loss reaches massive proportions, however, as is the case for many parts of the developing nervous system [Oppenheim, 1981], the occasional imprecision that gives rise to small-scale cell death is no longer a sufficient rationale for such a large-scale phenomenon and one must, instead, search for alternative explanations involving biological adaptiveness and survival value... The occurrence of massive cell death during ontogeny is not unique to neurons, but rather it appears to be a phenomenon that has evolved so as to comprise a normal part of the development of many different types of cells, tissues and organs..." (Oppenheim, 1985, p. 487).

Oppenheim refers to the adaptation anomaly and points out a possible explanation to the anomalous phenomenon which will be investigated shortly.

The adaptation anomaly of the nervous system can also be observed in mature organisms, for example, during the re-generation (pruning) of injured organs in species which are able to re-generate whole organs. A typical re-generation behaviour consists of a rapid proliferation of injured organ cells to the extent of excessive over-development, then followed by the typical bottle-neck behaviour.

Entropy, Information and Adaptation

In the previous section we learned that the adaptation anomaly (or bottle-neck phenomenon) is a 'popular' strategy in nature. This behaviour of an adaptive system will be formulated in terms of information content and fitness which may help us to understand why so many systems exhibit the adaptation anomaly in spite of its seemingly wasteful character. Entropy, a basic concept of thermodynamics, is used in this discussion in its almost literal meaning. It is necessary to consider a new entropy measure for information systems in spite of the considerable work already available on the adoption of entropy to that domain because most of the adopted definitions of entropy in information theory serve to describe only a partial and synthetic aspect of information, not appropriate for our purposes. It was already mentioned that for natural systems the whole is more than the sum of the parts. In natural systems, the information contained by the system cannot be separated from the system and the way this information is being used, and it cannot be separated from the environment.

Before proposing an analogy between entropy and information processing of natural adaptive systems, the meaning of entropy in thermodynamics is briefly considered. The first law of thermodynamics – the law of energy conservation – states that as a system undergoes a change of state, energy may cross the boundary either way, and that the net energy change will be exactly equal to the net energy that crossed the boundary of the system. Although the first law quantifies the state changes during any process, it does not specify the direction in which a process will occur. In the eyes of the first law, a cup of hot coffee can heat up further by absorption of heat from a cool surrounding just as it cools by virtue of heat transfer to it. It is clear from daily experience that some processes, like a cooling cup of coffee, can occur, while others cannot (such as the further heat absorption from a cool surrounding). The first law, however, does not indicate which processes are plausible and which are not. Experimental evidence led to the formulation of the second law of thermodynamics and to the consideration of an abstract thermodynamic property *entropy* which symbolizes, when loosely interpreted, the potential of a system to change or to undergo a thermal process. As a result of the second law, it is universally accepted that no process can occur if this process produces net entropy change (in net change one refers to a closed system). Putting it differently, the second law leads to the principle of the increase of entropy which states that no process can occur if it would result in a total decrease of entropy.

Another popular interpretation of entropy is that it indicates the amount of disorder in the system. The pictorial meaning of entropy as a measure of disorder is important for our discussion, and an illustrative paradigm of entropy in terms of disorder is presented. The paradigm is called: the freshman-graduate paradigm. A freshman's table is a good example of disorder (no offence, of course, to the rare exceptions). Due to several years of experience, the graduate's table looks very different. The freshman devotes excess effort to any study in which he is engaged because of the little space available (due to disorder). Many books are opened and more are piled, loose paper is scattered all over. Little additional disorder can be introduced to the table of our freshman since it already has so much. Often, the freshman realizes that too many notes and books are being used and there is absolutely no more room on his desk for additional work. At this stage, any new task (such as an unexpected exam) will require some tidying up before the new task can be tackled. In contrast to the freshman, the graduate (order trained by experience) can readily accommodate a fresh workload. If one considers the disorder aspect associated with entropy, one can say that the more orderly a system is the more flexible it is in adapting to new requirements. Entropy is an abstract concept that has been devised to aid the description of physical processes. We may conclude that the entropy change of a system is a measure of the system's ability to change further so that when the system's entropy decreases, it increases its potential to take part in more processes and therefore increases its adaptability. It is now simple to speak of the meaning of entropy in an adaptive information system.

Any information system can perform two basic operations: it can acquire or discard information, and/or it can manipulate the information it already has. In artificial information systems the two operations are technically unlimited (the cost of acquiring more information is often an artificial parameter). In natural systems however, the cost of any operation is a meaningful factor. In the case of a nervous system for example, an organism not only has to invest valuable time and energy to generate neural cells and connections, but it is also costly to maintain. Furthermore, the development of any organ has to be balanced and integrated with the functioning and available resource of the whole organism. It is clear that ultimately any growth or change in one organ affects the whole organism in terms of resources in a proportional fashion; the more complex an organ, the more demanding its needs for resources. The limited resources for growth and the cost of an organs' maintenance are poignant and intuitive aspects of adaptive natural systems. One can adopt this test of paucity or abundance as an entropy measure of the development state of an information system because it provides a measure of how much the system can evolve. The more developed an organ, relative to the resources available to its growth and functioning, the less potential it has to change or grow further. This is a good analogy to entropy in thermodynamics and proves to be effective for analyzing the possible development of a natural system as will be shown shortly.

Let us now return to the question of the advantage of increasing adaptability and the entropy changes involved. What follows from our analogy between entropy and natural systems is that relative growth indicates an increase in the entropy of the system and therefore indicates that the system is using more resources, and that it is closer to the limit of resources available to it. In terms of disorder and 'cumbersomeness' of a system, a relative growth increases the complexity of the system, and therefore entropy increases. This suggests that the system can undergo fewer changes in its present state. The same considerations apply when the system is simplified. Simplifying a system reduces the resources the system requires and thus allows the system greater resources for development and change. Thus far we have equated the available resources and complexity with entropy and the potential for growth, but we have not tied the two to adaptability and fitness. In terms of our original paradox, one needs to determine what natural system would gain by simplifying itself and thus increase its potential for development, only to grow again and lose that potential?

The answer to the question: "What fitness gain does an adaptive system achieve by reducing its entropy before increasing it again?", is simple when one considers the environment in which the adaptation occurs. Certain aspects of the environment cannot be predicted. This relative state of uncertainty cannot be coded in the organism, and therefore cannot be provided by ready made solutions realized by long-term evolution. In these situations, an organism which assumes little about the environment is more robust. Robustness means less specialization, but necessitates greater redundancy to maintain functionality. The organism has invested resources in redundancy in order to be fit for a wider range of environmental conditions. Once adaptation commences towards a specific environment, much of the redundancy of the 'prudent' organism can be discarded as it contributes little to its fitness. Furthermore, by simplifying itself the organism increases its available resources for growth, which are needed for its further specialization. A redundant system and a mechanism which allows the simplification of the system before specialization takes place, is a good strategy in situations where there is a degree of uncertainty towards what conditions the system will need to adapt to. It is a wasteful process to produce redundancy simply in order that it can be reduced in order to grow again, but it is in the long term a rewarding strategy which bestows higher fitness. It is now less surprising that such a strategy is so 'popular' in nature.

We return to the development of the nervous system during embryogenesis in terms of development, fitness and entropy [Cowan, 1979; Liu, 1981]. Initially, an organism develops redundancy in the system and possesses multiple neurons and connections which compete for control. This initial state is the reference state of entropy for the nervous system and the system is in relative disorder because the connections are largely made at random. Due to selection pressure and competition among the neurons during the embryogenetic period, a few selected neurons win, and the rest of the neurons (and connections) die. At this stage the entropy of the system is reduced because the system is much smaller and simpler. Fewer sensory signals can be transmitted and poorer sensory information is produced. The whole nervous system becomes a less complex system, and entropy of the system decreases. This indicates that the system has greater ability to develop relative to what it was before, and this is intuitively apparent. After a period of time, the remaining neurons begin to grow and to differentiate, demanding more resources to accomplish this task. At this stage, the nervous system becomes more complex and the entropy of the system increases. The nervous system becomes specialized, and attains a higher fitness, but loses its robustness. This is a perfect analogy to the adaptation anomaly of robot path-planning, and is also an explanation as to why it occurs. By exercising selection against neurons and connections, the nervous system gains a growth potential which it needs in order to specialize those neurons which were better fitted. The rationale behind the adaptation anomaly of the trajectory-GA will be discussed in the following section.

The Evolution of Trajectories

In this section, the adaptive behaviour of the trajectory-GA is analyzed in detail. An explanation for the adaptive anomaly will be proposed both in terms of the specific trajectory environment and in terms of the similar adaptation phenomena observed in nature.

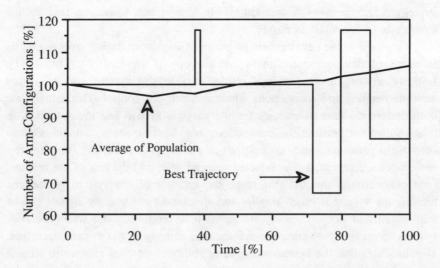

Fig. 64 - *The average number of arm-configurations and the number of arm-configurations in the best trajectory at a transitional stage of the search.*

'Unique' Reasons for the Anomaly in the Trajectory-GA

As presented earlier, the average number of arm-configurations during an optimization of a population of trajectories, initially decreases and later increases (Fig. 64). What causes this behaviour? The performance of a trajectory and consequently its fitness is affected by various factors which differ in importance. The performance of a trajectory is the sum effect of these factors, each one having its own weight (proportional contribution), and an absolute quality. The importance of the various fitness factors is not equal and is not even fixed. At different stages of the optimization, the trajectory-GA directs its processing resources to the different factors according to their weight. This dynamic shift in search effort is not explicitly controlled by the GA, but rather implicitly by the fitness function. The average number of arm-configurations demonstrates this dynamics.

During the initial stages of the search, when trajectories are formed from random collections of arm-configurations with no attention being paid either to the streamline aspect of the sequence of arm-configurations or to the end-effector positions, the main source for inadequate performance is the movement between unharmonious arm-configurations. A displacement vector which is composed of two considerably different arm-configurations (irrespective of the position of the end-effector like the illustrative motion in Fig. 26), will result in a vigorous end-effector movement which causes a large deviation from the specified desired path. At this stage, short trajectories with few arm-configurations have an advantage over trajectories with a large number of arm-configurations. This is because the probability of arm-configurations being co-adapted is inversely proportional to their number. Thus, during the initial stages of the search, it is expected that the average number of arm-configurations will tend to drop.

After a period of relaxation, during which the trajectories become harmonious (no vigorous arm displacements occur between adjacent end-effector positions) due to crossover and the emergence of co-adapted arm-configuration sequences, performance can further be improved in two ways:

(1) By optimizing the current arm-configurations within a trajectory, so that they will be more harmonious and accurate.
(2) By increasing the number of arm-configurations composing a trajectory.

As the search progresses, trajectories become more harmonious and the 'unharmoniousness' factor becomes less significant. Substantial improvements in performance can then be gained by increasing the accuracy of the trajectory. An increase in the number of arm-configurations is directly linked to the frequency with which the robot (and the simulation) motion controller can sample the path and the amount of feedback it receives about its progress (this sampling occurs at each discrete end-effector position in a trajectory). The accuracy of path-following is proportional to the amount of feedback and thus to the number of arm-configurations. At the stage when unharmoniousness and accuracy have an equal effect over performance, the average number of arm-configurations stabilizes, and subsequently the average number of arm-configurations increases to allow greater accuracy.

General Reasons for the Adaptation Anomaly

After the observations on adaptation anomaly in natural systems and following the discussion on entropy considerations in natural systems, the trajectory anomaly can be explained in a more general way. Adaptive systems which are subject to selection follow two main strategies. The first is to be specialized and incorporate as much functioning specific information as possible. Such systems are efficient, but not too flexible and therefore their adaptability is limited. The second strategy is to be very flexible, so that the system can adapt easily. The latter strategy is more wasteful than the former, but is a necessity when there is no information on the environment or when the environment changes drastically. The second strategy is the one in which we are interested, and is the strategy adopted by all the mechanisms that exhibit the adaptation anomaly. The adaptation anomaly now seems to be natural and logical. When the conditions in which adaptation will take place are not known, a strategy that leaves adaptation to the very last minute is robust. In order to adapt well, the adaptability of the system has to be optimized. The adaptability is optimized if more growth resources are supplied to the system. When a relative high state of redundancy is developed, selection can select those parts which are more fit and discard the rest, thus reducing the entropy of the system while at the same time improving the fitness. From the relatively low entropy state, the system has significantly more resources to improve the adaptation by further specialization. At this stage, the entropy of the system grows, but so does the fitness, which was the original and main object of the whole manœuvre.

The prerequisite features of a system needed in order that it can optimize its adaptability as part of its straightforward adaptation are:

(1) An initial creation of redundancy which can be reduced without major loss of functionality.
(2) The ability to simplify the system by selection.
(3) The ability to grow and specialize after simplification.

The above three conditions are implicitly provided by natural selection, provided the system can produce and function under redundant conditions. The trajectory-GA and the natural representation used for this model embody these features and therefore exhibit the bottle-neck behaviour.

Summary

This chapter postulated that general adaptive systems have a characteristic behaviour in which adaptation is achieved through the optimization of the system's ability to adapt. In order to optimize the ability to adapt, the system simplifies its information structure and organization. Natural systems achieve structural adaptation by manipulating the degree of redundancy in an indirect manner by way of natural selection operating on redundant parts of the system. A similar adaptive behaviour can be observed with a GA. This postulate is based on biological evidence on both micro- and macro-levels of evolution.

The conclusiveness of the biological evidence leaves little room for creative interpretation. It may be observed that adaptation of many natural systems characteristically starts from a relatively chaotic and redundant state so it can adapt to diverse conditions through the simplification and structuring of the system before specialization. This phenomenon was termed *adaptation anomaly*. This chapter also proposed an explanation of this phenomenon in terms of increased fitness due to robust adaptability. The explanation is based on strong analogies between adaptation and entropy, and on experimental results obtained from an artificial environment which reproduces the evolution anomaly of complexity growth. Although the adaptation anomaly is characteristic of the trajectory-GA and is repeatable under diverse test conditions, it is recognized that the robot model does not constitute a general model of natural adaptation and therefore one should be cautious in projecting its results on other adaptive systems. Nonetheless, the conclusiveness in which the negative entropy phenomenon presents itself in nature does provide powerful support for the proposed model of information processing through adaptation of natural systems.

What are the lessons one might draw from the evidence and the analysis provided about the adaptation anomaly mechanisms in nature? The most important one is that natural selection operates on all levels of life, both the micro and macro alike. The effect of selection on some levels is not seen sometimes due to the strong effect it has on other levels, but sure enough selection is there. Another important aspect of adaptation is the advantage redundant systems have in that they can afford to discard information and by that optimize their ability to adapt. Redundant representations in GAs were not popular so far due to theoretical and practical implications. However, the advantage encompassed by the redundant models is potentially rewarding to the extent that such a type of representation should be further investigated.

The author wishes to conclude this chapter with two remarks of a somewhat personal nature. One, that it is recognized that the ideas presented in this chapter are speculative to a certain extent. This was not intended. The main objective was to highlight the importance of allowing free rein for selection and adaptation when the objective is optimization of an adaptive process. Two, when purpose-dedicated GAs are constructed for particular problems, customizing the algorithm is justified. However, when GAs are researched with the intention to deepen the knowledge of adaptation in natural and artificial systems, it is vital to leave as much plasticity as possible to the algorithm to define its own optimum parameters through the exposure of all the GA components to selection.

Chapter 11
CONCLUDING REMARKS

This book grew out of a process control problem in robotics I found difficult to approach with classical optimization methods. Having no clear answer 'how' to solve the problem, I turned it around and asked 'why' such a problem is not amenable to traditional problem solving methods. A fellow student who noticed my anguish quoted a lecturer who said "...if you cannot answer the question, model the issue...". In as much as my interest in GAs was originally aimed at solving a specific problem, GAs, with their immense diversity and richness, steered my interests towards the general issues of adaptation in natural and artificial systems. What struck me most – and still does – is that in contrast to the relatively simplistic mechanisms of a GA, its behaviour as a whole is complex. This facet of complex systems, the inherent difference between the functioning of the whole and the sum functioning of the parts, is a fundamental aspect which explains the sophistication of GA applications and, of course, of nature. This is probably the reason why GAs are not as widely used as those acquainted with the basic theory might expect. In hindsight, I realize that I was far too pre-conditioned to regard sophistication as synonymous with 'intelligence' and intelligence with determinism. Nature seems to indicate the opposite. We shall probably never know whether determinism was properly tested by nature, and the fact that such mechanisms did not survive the course of evolution really means that they are less fit. What we do know however, and partly from research in GAs, is that much of the day to day mechanisms in nature are not deterministic. The mandate that was given to determinism by many artificial intelligence schools as the correct approach to explain nature's sophisticated mechanisms often proves to be limiting. Much of the amazing behaviour is a product of a relatively simple and implicit search process which is really 'blind'.

A bee searching for nectar exhibits a certain search behaviour – a strategy optimize her effort. Whatever her strategy is, whether it is a random search or a more structured one, this strategy did not come about through any analytical thought process. New strategies, or rather behaviours, surface through reproduction. Once different behaviours are present, selection steps in. Selection just simply 'happens', and its effect is determined according to the sum total fitness of the organism. Evolutionists may find it difficult to reconstruct the correct fitness function which acts on the bee, the 'complete picture' of what is important and what is not in the pursuit of nectar. However, that should not be interpreted as a difficulty which nature has in creating new strategies and in selecting the better ones. It is we that have a difficulty in deciphering what nature finds easy to code.

It is inevitable that an interdisciplinary work of this type should touch on many issues, and equally inevitable that some should be left unresolved. In a way, all the issues discussed in the book emerge from our early choice of representation, a flexible representation format which incorporated both over- and under-specified trajectories. No doubt it is the flexible representation structure which enabled the successful application of a GA to the domain of trajectory generation. The original objective in using such an unconventional representation structure was not motivated by oddity for the sake of oddity. Rather, it was the belief that such representations are the essence and a vital ingredient in the evolutionary process, which enables the mechanisms of evolution to be realized to their maximum. To end this book with a sentence which encapsulates the 'gospel' of the book, one has to resort to the much used and abused phrase "less is more". Or, in the immortal words of Jerome K. Jerome, "...we must not think of the things we could do with, but only of the things that we can't do without."

APPENDIX A

The Robot Dynamic Simulation Package ROSI

ROSI is a simulation package designed for the dynamic simulation of industrial robots. This package was designed by Dr Leon Zlajpah of Jozef Stefan Institute. Fig. A1 shows the structure of the package: manipulator, actuators, gears, sensors, control and trajectory generation.

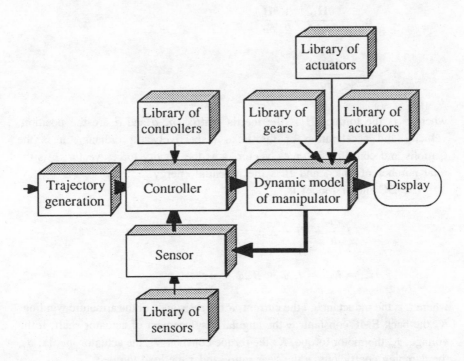

Fig. A1 *- The modules composing the ROSI robot dynamic simulation package.*

The various libraries include some basic models for each group of module, which allow the user to configure a wide range of robot system specifications. To create a simulation file, the relevant modules are linked together. The main part is the dynamic model of the manipulator and includes the kinematic, dynamic and mathematical model of the manipulator, the model of the actuators and the model of the gears. Using the Lagrange equation, the mathematical model of the manipulator must be given in a closed form and in the format

$$\tau = \mathbf{H}(q)\ddot{q} + \mathbf{h}(q, \dot{q}) + \mathbf{k}(q)$$

so for each link,

$$\tau_i = \sum_{j=1}^{n} \mathbf{H}_{ij}\ddot{q}_j + \sum_{j=1}^{n} \sum_{l=1}^{n} \mathbf{h}_{ijl}\dot{q}_j\dot{q}_l + \mathbf{k}_i$$

$$\mathbf{h}_{ijl} = \frac{\partial \mathbf{H}_{ij}}{\partial q_l} - \frac{1}{2} \frac{\partial \mathbf{H}_{jl}}{\partial q_i}$$

$$\mathbf{k}_i = \frac{\partial P(q)}{\partial q}$$

where τ is the torque, $\mathbf{H}$ is the inertia matrix, q, $\dot{q}$ and $\ddot{q}$ are the position, velocity and acceleration of the links respectively (in radians), $\mathbf{h}$ is the Coriolis and centrifugal forces vector, $\mathbf{k}$ is the gravity forces vector, n is the total number of links, and P is the potential energy.

The DC motor model is

$$L\frac{di}{dt} + Ri + K_b\dot{\varphi} = u$$

$$\tau_m = K_\tau i = I_m\ddot{\varphi} + B_m\dot{\varphi} + n\tau$$

where L is the inductance, i the current, R the resistance of the armature winding, K_b the back EMF constant, φ the angular displacement of actuator shaft, u the voltage, τ_m the motor torque, K_τ the torque constant, I_m the actuator inertia, B_m the damping coefficient, n the gear ratio, and τ the load torque.

Choosing

$$x_i = (q_i, \dot{q}_i)^T = n(\varphi_i, \dot{\varphi}_i)^T \ ,$$

the model is transformed to

$$\dot{\mathbf{x}}_i = \begin{bmatrix} 0 & 0 \\ 0 & -a_i \end{bmatrix} \mathbf{x}_i + \begin{bmatrix} 0 \\ b_i \end{bmatrix} \mathbf{u}_i + \begin{bmatrix} 0 \\ -g_i \end{bmatrix} \tau_i$$

where

$$a_i = \frac{RB_m + K_1 K_b}{RI_m}$$

$$b_i = \frac{nK_\tau}{RI_m}$$

$$g_i = \frac{n^2}{I_m} \ .$$

Combining the dynamic model of the manipulator and the models of actuator,

$$\dot{\mathbf{x}}_i = \begin{bmatrix} 0 & 0 \\ 0 & -a_i \end{bmatrix} \mathbf{x}_i + \begin{bmatrix} 0 \\ b_i \end{bmatrix} \mathbf{u}_i + \begin{bmatrix} 0 \\ -g_i \end{bmatrix} (\mathbf{H}(q)\ddot{q} + \mathbf{h}(q, \dot{q}) + \mathbf{k}(q))$$

where $\mathbf{A} = \text{diag}(a_i)$, $\mathbf{B} = \text{diag}(b_i)$, and $\mathbf{G} = \text{diag}(g_i)$.

The complete model of the robot can be represented in the form

$$\ddot{q} = (\mathbf{I} + \mathbf{GH}(q))^{-1}[-\mathbf{A}\dot{q} + \mathbf{B}\mathbf{u} - \mathbf{G}(\mathbf{h}(q, \dot{q}) + \mathbf{k}(q))] \ .$$

Link velocities and positions are generated from link acceleration using the Euler method of integration. To compensate for the simple integration method and to ensure accurate results, the sampling frequency for the calculation of the dynamic model is approximately 10 kHz.

BIBLIOGRAPHY

Ackley, D. H. (1985). A connectionist algorithm for genetic search. *1st International Conference on Genetic Algorithms*, 121-135.

Asada, H. and Slotine, J.-J. E. (1986). *Robot analysis and control*. John Wiley.

Bagley, J. D. (1967). The behaviour of adaptive systems which employ genetic and correlation algorithms (Doctoral dissertation, University of Michigan). *Dissertation Abstract International, 28(12), 5106B.*

Berek, C., Griffiths, G. M. and Milstein, C. (1985). Molecular events during maturation of the immune response to oxazolone. *Nature*, 316, 412-418.

Bethke, A. D. (1980). Genetic algorithms as function optimizers (Doctoral dissertation, University of Michigan). *Dissertation Abstract International, 41(9), 3503B.*

Bickel, A. S. and Bickel, R. W. (1987). Tree structured rules in genetic algorithms. *2nd International Conference on Genetic Algorithms*, 77-81.

Booker, L. B. (1982). Intelligent behavior as an adaption to the task environment (Doctoral dissertation, University of Michigan). *Dissertation Abstracts International, 43(2), 469B.*

Bridges, C. L. and Goldberg, D. E. (1989). *A note on the non-uniform Walsh-schema transform* (TCGA Report No. 89004). Tuscaloosa: University of Alabama.

Cavicchio (1970). Adaptive search using simulated evolution. Unpublished Doctoral dissertation, University of Michigan.

Cohoon, J. P., *et al.* (1987). Punctuated equilibria: A parallel genetic algorithm. *2nd International Conference on Genetic Algorithms*, 148-154.

Cowan, W. M. (1979). The development of the brain. *Scientific American*, 241(3), 106-117.

153

Craig, J. J. (1986). *Introduction to robotics: Mechanics and control.* Reading: Addison-Wesley.

Crow, J. F. and Kimura, M. (1970). *An introduction to population genetics theory.* New York: Harper and Row.

Davidor, Y. and Davies, B. L. (1987). Criteria for robot performance in path-following activities. *1st International Conference on the Robotics.*

Davidor, Y. and Davies, B. L. (1988). A vision system for the on-line quality control of adhesive beads. *International Journal of Adhesion and Adhesives*, 8(1), 33-38.

Davidor, Y. and Davies, B. L. (1989a). Criteria for robot performance in path-following activities. *Robotics & Computer-Integrated Manufacturing*, 5(2/3), 191198.

Davidor, Y. and Dean, M. (Eds.), (1989b). *CROBAS Research Report.* London: Imperial College, University of London.

Davidor, Y. (1989c). Analogous crossover. *3rd International Conference on Genetic Algorithms*, 98-103.

Davidor, Y. (1989d). *Genetic algorithms for order dependent processes applied to robot path-planning.* (Unpublished Doctoral dissertation, Imperial College, University of London).

Davidor, Y. (1990). Robot programming with a genetic algorithm. *IEEE International Conference on Computer Systems and Software Engineering.*

Davidor, Y, (in press). A genetic algorithm applied to robot path-planning. In D. Davis (Ed.), *The Genetic Algorithm Handbook.* Van Nostrand Reinhold.

Davidor, Y., Jones, A. J. and Husband, T. (in press). Genetic algorithms for autonomous robot programming. In K. Warwick (Ed.), *Robotics: Applied mathematics and computational aspects.* Oxford University Press.

Davis, L. (1985). Job shop scheduling with genetic algorithms. *1st International Conference on Genetic Algorithms*, 136-140.

Davis, L. and Coombs, S. (1987). Genetic algorithms and communication link speed design: Theoretical considerations. *2nd International Conference on Genetic Algorithms*, 252-256.

Deb, K. and Goldberg, D. E. (1989). An investigation of niche and species formation in genetic function optimization. *3rd International Conference on Genetic Algorithms*, 42-50.

De Jong, K. (1975). An analysis of the behavior of a class of genetic adaptive systems (Doctoral dissertation, University of Michigan). *Dissertation Abstracts International 36(10), 5140B*.

Dolan, C. P. and Dyer, M. G. (1987). Towards the evolution of symbols. *2nd International Conference on Genetic Algorithms*, 123-131.

Englander, A. C. (1985). Machine learning of visual recognition using genetic algorithms. *1st International Conference on Genetic Algorithms*, 197- 202.

Fourman, M. P. (1985). Compaction of symbolic layout using genetic algorithms. *1st International Conference on Genetic Algorithms*, 141- 153.

Glover, D. E. (1987). Solving a complex keyboard configuration problem through generalized adaptive search. In L. Davis (Ed.), *Genetic algorithms and simulated annealing* (pp. 12-31). London: Pitman.

Goldberg, D. E. (1983). Computer-aided gas pipeline operation using genetic algorithms and rule learning (Doctoral dissertation, University of Michigan). *Dissertation Abstracts International, 44(10), 3174B*.

Goldberg, D. E. (1985). *Optimal initial population size for binary-coded genetic algorithms* (TCGA Report No. 85001). Tuscaloosa: University of Alabama.

Goldberg, D. E. (1987a). *A note on the disruption due to crossover in a binary coded genetic algorithm* (TCGA Report No. 87001). Tuscaloosa: University of Alabama.

Goldberg, D. E. (1987b). Simple genetic algorithms and the minimal, deceptive problem. In L. Davis (Ed.), *Genetic algorithms and simulated annealing* (74-88). London: Pitman.

Goldberg, D. E. (1988). *Genetic algorithms and Walsh functions: Part I, a gentle introduction* (TCGA Report No. 88006). Tuscaloosa: University of Alabama.

Goldberg, D. E. (1989a). *Genetic algorithms and Walsh functions: Part II, deception and its analysis* (TCGA Report No. 89001). Tuscaloosa: University of Alabama.

Goldberg, D. E. (1989b). *Genetic algorithms in search, optimization & machine learning*. Reading: Addison-Wesley.

Goldberg, D. E., (1989c). Real alphabets. Personal communication.

Goldberg, D. E. (1989d). Zen and the art of genetic algorithms. *3rd International Conference on Genetic Algorithms*, 80-85.

Goldberg, D. E., Korb, B. and Deb, K. (1989e). *Messy genetic algorithms: Motivation, analysis, and first results* (TCGA Report No. 89003). Department of Engineering Mechanics, Tuscaloosa: University of Alabama.

Goldberg, D. E. and Richardson, J. (1987). Genetic algorithms with sharing for multimodal function optimization. *2nd International Conference on Genetic Algorithms*, 41-49.

Goldberg, D. E. and Samtani, M. P. (1986). Engineering optimization via genetic algorithm. *Ninth Conference on Electronic Computation*, 471- 482.

Goldberg, D. E. and Smith, R. E. (1987). Nonstationary function optimization using genetic algorithms with dominance and diploidy. *2nd International Conference on Genetic Algorithms*, 59-68.

Gorczynski, R. M. and Steele, E. J. (1981). Simultaneous yet independent inheritance of somatically acquired tolerance to two distinct H-2 antigenic haplotype determinants in mice. *Nature*, 289, 678-681.

Greene, D. P. and Smith, S. F. (1987). A genetic system for learning models of consumer choice. *2nd International Conference on Genetic Algorithms*, 217-223.

Grefenstette, J., (1979). *Representational dependencies in genetic algorithms.* Unpublished manuscript.

Grefenstette, J. J. (1981). *Parallel adaptive algorithms for function optimization* (CS-81-19). Computer Science Department, Vanderbilt University.

Grefenstette, J. J. and Fitzpatrick, J. M. (1985). Genetic search with approximate function evaluations. *1st International Conference on Genetic Algorithms*, 112-120.

Haralick, R. M., Sternberg, S. R. and Zhuang, X. (1987). Image analysis using mathematical morphology. *IEEE Transactions on Pattern Analysis and Machine Intelligence*, 9(4), 532-550.

Hatvany, J., (1987). What makes robots tick?. 1st International Conference on the Robotics, Dubrovnik, Yugoslavia.

Holland, J. H. (1971). Processing and processors for schemata. In E. L. Jacks (Ed.), *Associative information processing* (pp. 127-146). New York: American Elsevier.

Holland, J. H. (1973). Schemata and intrinsically parallel adaptation. *Proceedings of the NSF Workshop on Learning System Theory and its Applications*, 43-46.

Holland, J. H. (1975). *Adaptation in natural and artificial systems*. Ann Arbor: University of Michigan Press.

Holland, J. H. and Reitman, J. S. (1978). Cognitive systems based on adaptive algorithms. In D. A. Waterman and F. Hayes-Roth (Ed.), *Pattern directed inference systems* (313-329). New York: Academic Press.

Hollstien, R. B. (1971). Artificial genetic adaptation in computer control systems (Doctoral dissertation, University of Michigan). *Dissertation Abstracts International, 32(3), 1510B*.

Jacobson, H. (1955). Information reproduction and the origin of life. *American Scientist*, 43(1), 119-127.

Jog, P. and Van Gucht, D. (1987). Parallelisation of probabilistic sequential search algorithms. *2nd International Conference on Genetic Algorithms*, 170-176.

Jones, A. J., (1988). *Improving the reliability of genetic algorithms when searching for an optimal solution in large multi-modal spaces*. Unpublished manuscript.

Jordan, M. I. and Rosenbaum, D. A. (1988). Action. In M. L. Posner (Ed.), *Handbook of Cognitive Science*. Cambridge: MIT Press.

Kandel, E. R. (1979). Small systems of Neurons. *Scientific American*, 241(3), 60-70.

Khoogar, A. R., (1987). Genetic algorithm solutions for inverse robot kinematics (TCGA File No. 01252). Tuscaloosa: University of Alabama.

Lenarcic, J., Stanic, U. and Oblak, P. (1987). Some kinematic considerations for the design of robot manipulators. *1st International Conference on the Robotics*,

Liu, M. (1981). *Biology and pathology of nerve growth*. New York: Academic Press.

Mars, P. (1989). Neural networks and robotic control. *IMA National Conference on Robotics: Applied Mathematics and Computational Aspects*.

Maynard-Smith, J. (1989). *Evolutionary genetics*. Oxford University Press.

Oppenheim, R. (1981). In W. M. Cowan (Ed.), *Studies in developmental neurobiology* (pp. 74-133). Oxford University Press.

Oppenheim, R. W. (1985). Naturally occurring cell death during neural development. *Trends in Neuroscience*, 8, 487-493.

Parker, J. K., Khoogar, A. R., and Goldberg, D. E. (1989). Inverse kinematics of redundant robots using genetic algorithms. *IEEE International Conference on Robotics and Automation*, Vol. 1, 271-276.

Paul, R. P. and Zong, H. (1984). Robot motion trajectory specification and generation. *2nd International Symposium on Robotics Research*.

Platt, J. R. (1961). Properties of large molecules that go beyond the properties of thier chemical sub-group. *Journal of Theoretical Biology*, 1, 342-358.

Ptashne, M. (1989). How gene activators work. *Scientific American*, 25-31.

Rosenberg, R. S. (1967). Simulation of genetic populations with biochemical properties (Doctoral dissertation, University of Michigan). *Dissertation Abstracts International, 28(7), 2732B*.

Rumelhart, D. E., Hinton, G. E. and Williams, R. J. (1986). Learning representations by backpropagating errors. *Nature*, 323, 533-536.

Sagi, D. and Julesz, B. (1987). Short-range limitation on detection of feature differences. *Spatial Vision*, 2(1), 39-49.

Samuel, A. L. (1959). Some studies in machine learning using the game of checkers. *IBM Journal of Research and Development*, 3, 211-223.

Schaffer, J. D. (1984). Some experiments in machine learning using vector evaluated genetic algorithms. Unpublished Doctoral dissertation, Vanderbilt University.

Schaffer, J. D. (1985a). Learning multiclass pattern discrimination. *1st International Conference on Genetic Algorithms*, 74-79.

Schaffer, J. D. (1985b). Multiple objective optimization with vector evaluated genetic algorithms. *1st International Conference on Genetic Algorithms*, 93-100.

Schaffer, J. D. (1987). An adaptive crossover distribution mechanism for genetic algorithms. *2nd International Conference on Genetic Algorithms*, 36-40.

Schaffer, J. D. and Grefenstette, J. J., (1988). *A critical review of genetic algorithms*. Unpublished Technical Report. Philips Laboratories, Briarcliff Manor, NY.

Shaefer, C. G. (1987). The ARGOT strategy: Adaptive representation genetic optimizer technique. *2nd International Conference on Genetic Algorithms*, 50-58.

Shaefer, C. G. (1988). The ARGOT strategy: Combinatorial optimizations *International Workshop on Change of Representation and Inductive Bias*.

Simon, H. A. (1962). The architecture of complexity. *Proceedings of the American Philosophical Society*, 106(6), 467-482.

Smith, S. F. (1980). *A learning system based on genetic adaptive algorithms*. Unpublished Doctoral dissertation, University of Pittsburgh.

Smith, S. F. (1984). Adaptive learning systems. In R. Forsyth (Ed.), *Expert Systems, Principles and Case Studies*. London: Chapman and Hall.

Stadnyk, I. (1987). Schema recombination in a pattern recognition problem *2nd International Conference on Genetic Algorithms*, 27-35.

Suh, J. Y. and Gucht, D. V. (1987). Incorporating heuristic information into genetic search. *2nd International Conference on Genetic Algorithms*, 100-107.

Syswerda, G. (1989). Uniform crossover in genetic algorithms. *3rd International Conference on Genetic Algorithms*, 2-9.

Tanese, R. (1987). Parallel genetic algorithm for a hypercube. *2nd International Conference on Genetic Algorithms*, 177-184.

Tsotsos, J. K. (1987). A 'complexity level' analysis of vision. *1st International Conference on Computer Vision*, 346-355.

Vintsyuk, T. K. (1971). Element by element recognition of continuous speech composed recognition. *Proceedings of International Conference of Acoustics*, 133-143.

Weismann, A. (1904). *On evolution*. London: Edward Arnold.

Westerdale, T., (1989). *Nonlinear scaling*. Personal comunication.

Wilson, S. W. (1985). Adaptive "cortical" pattern recognition. *1st International Conference on Genetic Algorithms*, 188-196.

Wilson, S. W. (1987). Hierarchical credit allocation in a classifier system. In L. Davis (Ed.), *Genetic algorithms and simulated annealing* (pp. 104-115). London: Pitman.

Winfield, D. A. (1983). The postnatal development of synapses in the different laminae of the visual cortex in the normal kitten and in kittens with eyelid suture. *Developmental Brain Research*, 9, 155-169.

Winston, P. H. (1981). *Learning new principles from precedents and exercises: The details* (AIM 632). Massachusetts Institute of Technology.

Winston, P. H. (1984). *Learning by augmenting rules and accumulating censors* (AIM 678). Massachusetts Institute of Technology.

INDEX